EXPERIMENTING WITH HUMANS AND ANIMALS

JOHNS HOPKINS

INTRODUCTORY STUDIES

IN THE HISTORY

OF SCIENCE

Experimenting with Humans and Animals

From Aristotle to CRISPR

Second Edition

Anita Guerrini

JOHNS HOPKINS UNIVERSITY PRESS

BALTIMORE

© 2003, 2022 Johns Hopkins University Press
All rights reserved. Published 2022
Printed in the United States of America on acid-free paper
9 8 7 6 5 4 3 2 1

The first edition of this work was published as *Experimenting with Humans and Animals: From Galen to Animal Rights.*

Johns Hopkins University Press
2715 North Charles Street
Baltimore, Maryland 21218-4363
www.press.jhu.edu

Library of Congress Cataloging-in-Publication Data

Names: Guerrini, Anita, 1953– author.
Title: Experimenting with humans and animals : from Aristotle to CRISPR / Anita Guerrini.
Description: Second edition. | Baltimore : Johns Hopkins University Press, 2022. | Series: Johns Hopkins introductory studies in the history of science | Includes bibliographical references and index.
Identifiers: LCCN 2021048341 | ISBN 9781421444055 (paperback) | ISBN 9781421444062 (ebook)
Subjects: LCSH: Vivisection—History. | Animal experimentation—History. | Human experimentation in medicine—History.
Classification: LCC QP45 .G84 2022 | DDC 179/.4—dc23/eng/20211001
LC record available at https://lccn.loc.gov/2021048341
A catalog record for this book is available from the British Library.

Special discounts are available for bulk purchases of this book. For more information, please contact Special Sales at specialsales@jh.edu.

Contents

Acknowledgments

The first edition of this book appeared in 2003, and a second printing, which corrected a few minor errors, appeared in 2009. A lot has happened in the life sciences since then. The COVID-19 lockdown in 2020 seemed like a good time to tackle an updated edition, and I am happy that Matt McAdam at Johns Hopkins University Press agreed. Matt's predecessor, Robert J. Brugger, suggested that I write about both animals and humans, and he was of course correct (as he was in many things pertaining to this book). Matt and his assistants, Will Krause and Adriahna Conway, answered many questions under trying conditions.

In writing this book, I have relied on the work of many scholars. Some of their works are listed in the Suggested Further Reading. The late Larry Badash first suggested I write about this topic, and Paul Farber, Diane McClure, Michael Osborne, Peter Neushul, and the series editors Mott Greene and Sharon Kingsland were instrumental is getting the first edition in shape. Many friends and colleagues, including Teresa Burgess, John Gluck, Peter McDermott, Marcia Meldrum, Karen Rader, Lindsey Reed, Jo-Ann Shelton, and Peter Sobol, read portions of the first edition. For the second edition, I am grateful to George Estreich and Michael Osborne for their careful reading of the entire text and to Nathaniel Comfort for reading chapter 7. Many conversations with Karen Rader over the years have helped to clarify my thoughts. At a summer workshop in 2003, "Ethical, Legal, and Social Implications of the Human Genome Project," I met Adrienne Asch, whose work has greatly influenced my thinking on disability. She is sorely missed. More recently, conversations with George Estreich have continued my education. Of course, none of these people are responsible for errors or misguided opinions. Many thanks to the students at the University of California, Santa Barbara, and at Oregon State University who took my class Animals in Science over the years and asked hard questions. The members and administrators of the Institutional Animal Care and Use Committees at UCSB and OSU have taught me much about regulation and oversight. I am grateful to the organizers of the 2018 Woods Hole/ASU History of Biology seminar "Engineering Life" for inviting me to participate and thus inspiring this new edition.

This book incorporates portions of the following works, in different form:

"A Tale of Two Rats," in *Nature Remade: Engineering Life, Envisioning Worlds*, ed. Luis Campos, Michael R. Dietrich, Tiago Saraiva, and Christian Young (University of Chicago Press, 2021), 31–43.

"The Human Experimental Subject," in *A Companion to the History of Science*, ed. Bernard Lightman (Wiley, 2016), 126–38.

"Animals and Ecological Science," in *The Oxford Handbook of Animal Studies*, ed. Linda Kalof (Oxford University Press, 2016), 489–505.

Most of all, I am grateful to my family, to my husband, Michael Osborne, our sons, Paul and Henry, and our daughter-in-law, Michelle, for Zoom cocktails and much more.

Introduction

How do we balance human needs and desires for drugs and vaccines, product safety, and medical advances, as well as our continuing thirst for knowledge about the natural world, with our responsibility to our fellow beings, human and nonhuman alike? The life sciences, including medicine, have relied on animal and human experimentation since at least the seventeenth century. When I began to write this edition in summer 2020, scientists were frantically searching for a vaccine against COVID-19. Employing techniques dating back to Pasteur in the nineteenth century, they continue to experiment with both animals and humans to find therapies and vaccines for this deadly disease. An article in *Science* magazine in April 2020 reported that "animals are studied for coronavirus answers," listing mice, hamsters, ferrets, and monkeys as possible candidates.[1] A week later *Nature* reported, "Hundreds of people volunteer to be infected with coronavirus" for a vaccine trial, even though no vaccine had yet been developed.[2]

Experimentation on both animals and humans has taken place nearly as long as Western science has existed, and these two areas of experimentation have always been closely intertwined. While most historians have treated them separately, I have chosen to look at them together. This is therefore a history of both humans and animals, but it is mainly a history of how humans have conceived of and used animals rather than a consideration of animals in themselves. Like the first edition, this book makes no claim to be comprehensive. Rather, it tracks the long history of human and animal experimentation through episodes that represent significant ideas or practices. Each chapter highlights a few topics and figures. Sidebars add case studies or give in-depth snapshots of subjects mentioned in the chapter. Although ethical considerations often enter the discussion, this book is not a history of biomedical ethics. It is a global history in the sense that Western-style science as defined by Europe and its diaspora has become the global standard for the production of knowledge about the natural world. A quick search through *Science* or *Nature* reveals the global reach of Western science in contributions from around the world that employ the same practices of experimentation, quantification, observation, and proof. I focus on practices in western Europe and North America, and much of my discussion in later chapters centers on the United States, which continues to dominate biomedical research. At the same time, ecology and other field sciences do not strictly adhere to the methodological models of the laboratory sciences, and this edition affords these sciences consideration as well.

Scientific practices, as historians of science have shown, are historical artifacts that have changed over time and from place to place. This book sketches a broad history of practices as employed in biology, physiology, medicine, and many subspecialties and therefore also offers a history of the development of these sciences as experiment-based disciplines. This edition is a thorough revision of the first edition, with many sections completely rewritten. Some sidebars have been deleted and others added. I have included much new material, reflecting the outpouring of work in the history of science, the history of medicine, and animal studies since 2000. An entirely new chapter, chapter 7, includes a discussion of the genomic revolution of the past two decades and how it has transformed animal and human experimentation. The Suggested Further Reading has also been updated to reflect recent scholarship.

The definition of experimentation varies widely according to time and place and includes external observation and manipulation; the testing of toxins in poisons, foods, or diseases; and vivisection, the surgical cutting open of the body, among other interventions in a body's normal functioning. Defined broadly, experimentation can also include the controlled observation that takes place in many field-based sciences.

Although confidence in the experiment dates from Western antiquity, we might describe the work of the ancients as *demonstrations* rather than experiments; instead of aiming to discover new facts about nature, they were more often intended to illustrate received beliefs. As defined in modern science, an experiment tests a hypothesis, a question about nature, under controlled conditions. Good researchers design an experiment so that it answers the question truthfully, that is, according to how nature really acts and not necessarily as the observer would want it to act. The investigator controls the setting of the experiment in order to account strictly for all the factors at play and to be reasonably certain of what cause in the experiment led to what effect. This often requires much trial and error, but failed experiments, in which nature does not act as the hypothesis predicted, can be as valuable as successful ones.

Experimenters therefore monitor physical variables such as temperature and air quality, use precise instruments, and carefully select subjects. Other variables that owe to the inherent unpredictability of the phenomena of life may be less easily managed, although, as we will see, standardization of animal subjects, particularly mice and rats, has eliminated at least some of this variation. The use of a control group that does not experience the experimental procedures is common in many biological experiments and clinical trials. These subjects are monitored for purposes of comparison. In experiments to test medicines, placebos—pills or treatments that do not contain the experimental element—perform the same function. In clinical trials of medicines on humans, the experiment is controlled for any possible psychological bias through the use of *double-blind* methods, in which neither the experimenter nor the experimental subject is aware of which substances are experimental.

This approach to experimentation owes much to practices, findings, and ideas of scientific professionalism that date from the late nineteenth and twentieth centuries. Although as early as the eighteenth century statistics helped to make the case for smallpox inoculation, statistical techniques became increasingly important in the twentieth century as researchers sought to minimize the influence of their preferences and also take into account the detailed characteristics of the subjects they studied. Experiments of any kind yield information, or data, which in recent experience usually appears in the form of quantitative statements of number, proportion, or statistical significance, and that information is then further manipulated using statistical techniques to make explicit the experimental outcomes.

Although observers often refer to experimentation on living beings (especially on animals) as *vivisection*, that term, strictly defined, means the cutting open of the animal body. Many experiments on animals and humans do not involve vivisection. In the pages that follow, the distinction between vivisection and other forms of experimentation is rendered as clearly as possible, but historically commentators did not always employ precise definitions.

Experimenting with animals and humans has a long history. The book begins in ancient Alexandria, some 2,300 years ago, when adventurous physicians took advantage of a (temporary) lack of customary taboos to engage in experimentation not only on dead animals and humans but on living ones. The Alexandrians' ideas about life derived from the Greek philosopher Aristotle. Aristotle's assumption of an analogy between animals and humans was taken up by the Roman physician Galen, who based his concepts of human anatomy on his dissection and vivisection of animals. Aristotle's and Galen's ideas on animals and humans dominated Western thinking until the seventeenth century. Early Christian theologians, following Aristotle, believed that while humans (possessed of reason and an immortal soul) existed for their own sake, animals (lacking reason) existed to serve human needs, establishing long-lived ideas of the relative place of humans and animals in nature—and therefore about experimentation.

In chapter 2 the story moves forward to the seventeenth century, focusing on the English physician William Harvey and his discovery of the circulation of the blood, published in 1628. Building on a century of anatomical work and on his own experimenting over the course of more than two decades, Harvey overthrew a central concept of Galen's theory of the workings of the body. For the remainder of the seventeenth century and into the eighteenth, researchers employed Harvey's techniques of dissection and vivisection to explore the implications of circulation. Many of these researchers followed the French philosopher René Descartes and believed that both the human and the animal body functioned as a machine. But most did not follow Descartes in his belief that animals, being only machines, were not conscious and therefore did not suffer pain as humans did.

The early-eighteenth-century controversy over smallpox inoculation, discussed in chapter 3, raised other issues concerning both humans and animals that remain with us today. Some physicians at the time claimed that inoculation—giving a mild version of the disease to a potential victim—conferred immunity to smallpox. But no one could explain how or why. Inoculation therefore remained experimental and risky. Many people, greatly fearing smallpox, accepted the risks of being inoculated. Comparing the survival rates of persons who were inoculated with those of persons who were not argued on behalf of the procedure. Yet the basic question of how much risk was acceptable remained unanswered. The discovery at the end of a century that a related animal disease, cowpox, could confer immunity to smallpox upended ideas about human and animal diseases and contributed to the development of veterinary medicine as a distinct field. At the same time, Enlightenment sensibilities led to new attention to experimentation on animals.

Experimental physiology became an essential science in the nineteenth century. As detailed in chapter 4, scientists experimented on many animals, particularly dogs, to penetrate the mysteries of human and animal function. The undoubted cruelty of many of these experiments led to a backlash, particularly after anesthesia was introduced around 1850 and yet was not always administered. An antivivisection movement, centered in Britain, led to passage by the British Parliament of the Cruelty to Animals Act of 1876, the first act to regulate animal experimentation. Women played a prominent role in this movement, in Britain and elsewhere.

At the time of the Cruelty to Animals Act, scientists could point to vastly increased knowledge of the human body as a justification for continued experimentation on animals. But their critics pointed out that this knowledge had thus far had few practical implications for human and animal health. As chapter 5 shows, this soon changed. The advent of the germ theory of disease, beginning in the 1870s, led to the development of vaccines and cures for many infectious diseases over the next 75 years. Developing vaccines and, later, antibiotics required thousands of animals, including rats, mice, dogs, rabbits, guinea pigs, and horses. In the 1880s, Louis Pasteur used hundreds of rabbits for his rabies vaccine; two decades later, Paul Ehrlich used thousands of mice to find Salvarsan, the first specific medicine to treat syphilis. Penicillin, discovered in 1928 but only operationalized in 1940, opened the door for dozens of potential new antibiotics, requiring the use of millions of animals to test efficacy and safety under new government guidelines. The antivivisection movement lost ground in this period as public opinion, bolstered by popular books and movies, shifted toward support of scientists.

Chapter 6 focuses on polio, a disease that in the twentieth century became a uniquely American story. Polio was an endemic disease of early childhood that, because of public health measures that had helped to stem cholera and typhoid, became an epidemic disease mainly of school-age children in the early twentieth century. In its most fearful manifestation, the poliovirus attacked the central nervous system and caused serious disabilities and even death. The story of polio encapsulates many of the themes introduced in previous chapters: the new experimental microbiology, complicated by the fact that polio was caused by a virus invisible to optical microscopes; the search for a vaccine; the role of institutions and funding; and the role played by clinical trials, first on vulnerable populations, and then a massive trial on children.

A significant part of this story is the major role played by primates, especially rhesus macaque monkeys, both in initial research on the disease and in the development of a vaccine. Galen had explained human anatomy using the body of an ape, and researchers returned to apes and monkeys at the end of the nineteenth century. Primates became important human surrogates in studies of disease and behavior, while field scientists such as Jane Goodall found them fascinating in themselves.

Chapter 7 begins at the end of World War II. The postwar years saw enormous growth in biomedical research, but regulation of that research as applied to both human and animal subjects proceeded slowly. New kinds of research emerged alongside traditional laboratory-based disciplines, in fields including ecology, ethology, and agriculture. From the 1990s onward, the new genetics profoundly affected research both in humans and in animals. The future of human and animal research is at a turning point, and we do not yet know the outcomes. The book returns in conclusion to the search for a COVID vaccine and how that search resonates with themes expressed in the previous chapters. The conclusion also looks ahead to new alternatives to traditional experimental animals, and to the One Health movement, which unifies human and animal medicine.

My purpose in writing this history is not simply to reveal abuses or to revel in the triumphs of science. I wish the story were that straightforward, because it would have been much easier to write. The story instead is one of trial and error, prejudice and leaps of faith, clashing egos and budget battles. I do not aim to designate certain historical players as right and others as wrong or to single out certain ideas as winners and others as losers. As a historian of science, I have spent much of my career examining "losing" ideas, not only for their intrinsic interest as intellectual problems but for what they reveal about human personalities and events. In addition, a close examination of almost any historical idea or scientific problem reveals unforeseen complications and contexts. Few things are simply right or wrong, either ethically or scientifically. More often they are a muddle of mixed motives and half-clear ideas.

The overriding theme of this book is not the conflict between science and ethics, as might be expected, but the interaction between science and society. Western-style life science has been remarkably successful in finding new facts about the world and in improving human and animal health. The costs have been high, to be sure, in both human and animal lives. It appears that most people, when they think about it, believe that the benefits outweigh the costs, although opinions vary widely. It is unlikely that these practices will disappear anytime soon. However, as readers will find in this book, regulation, often owing to public pressure, has significantly altered how science is conducted, mostly to the benefit of humans and animals as experimental subjects. By tracking the costs and benefits of human and animal experimentation through history, this book affords the reader an opportunity to assess the value, and values, of Western science. As the historian Susan Reverby writes, "Poorly done and misunderstood history is also a poor guide to policy."[3] Understanding the past enables us to influence the future.

1 Bodies of Evidence

Experimentation and Philosophical Debate in Premodern Europe

Alexandria in 280 BCE was a young city, vibrant and restless, its inhabitants coming from every corner of Alexander the Great's now-shattered empire. Alexander (356–323 BCE) had founded the city in 331 BCE as a cosmopolitan center of learning. Its academy, library, and medical school would be dominant forces in Western culture for a thousand years. In Alexandria, Greek culture thinly overlay a multitude of languages, religions, and ideas. Its combination of willing physicians, ambitious rulers, and what the historian Heinrich von Staden has called "scientific frontiersmanship"[1] allowed the dismissal of old taboos, at least for a time, including Greek prohibitions of the mutilation of the human body, living or dead.

Nonetheless, murmurs must have arisen in the marketplace at rumors that the king had given permission to the Greek physician Herophilus (ca. 330–ca. 260 BCE) and his younger colleague Erasistratus (304–245 BCE) to cut into a living man, not to cure him but merely to look inside. The man was a condemned criminal, scheduled for execution. Arguments for and against were heated. No contemporary accounts survive, but we hear faint echoes of these arguments in *De medicina*, written three centuries later by the Roman historian Celsus (ca. 25 BCE–ca. 50 CE).

Many medical sects flourished in Alexandria, but the opinions of two of these, the empirics and the dogmatists, are especially relevant. The empirical medics who set up practice in stalls in the marketplace were strongly opposed to any kind of dissection. Unlike the so-called philosophical physicians, who declared that medicine required knowledge of the natural world and how it worked, the empirics opposed any kind of theory. They relied on observation of the patient, claiming that philosophy and experimentation were irrelevant to medical practice. The physician could gain knowledge of diseases only through repeated observation. The ancient physician Hippocrates (ca. 460–ca. 375 BCE), in the casebooks known as the *Epidem-*

ics, supported this view with page after page of symptoms and signs and very little theorizing. Dissection and vivisection, said the empirics, gave knowledge of the dead, not of the living. In vivisection, they argued, the subject died anyway in the course of the operation, and the act of vivisection itself caused pathological changes that brought the validity of the observations into question. Observation of wounds during treatment could give much the same information without deliberately injuring a fellow human being. The empirics objected to vivisection on both intellectual grounds and moral ones, because the physician, as healer, should not cause suffering and death.

According to Celsus, Herophilus and Erasistratus were members of another sect known as the dogmatists. They believed that knowledge of anatomy was critical to medical practice, and they dissected to learn more about the body's internal workings, because mere observation of its exterior, even if supplemented by glimpses of wounds, was inadequate. In this they followed the model of the Greek philosopher and naturalist Aristotle (384–322 BCE), who had been Alexander the Great's tutor. Although Aristotle's goal was to learn more about how the human body works, he had limited access to the human body itself. Strong taboos against the mutilation of the human body existed in many ancient cultures, including ancient Greece. Surgery in the modern, invasive sense was unknown in antiquity—both because of these taboos and because of risks of infection and shock—and surgeons mainly treated wounds and set bones. Elaborate embalming and mummification rituals in ancient Egypt involved opening the dead body and removing its organs, but Egyptian physicians too were forbidden to cut open patients. In Greece, most regarded mutilation of the dead body with horror, seeing it as an act reserved for one's direst enemies in the course of war. In the tragedy *Antigone*, the title character goes to great lengths to ensure the proper burial of her brother's body. For Aristotle and the dogmatists, then, dissecting and experimenting on animals were necessary alternatives to research on humans.

Herophilus and Erasistratus dissected and vivisected animals, but although they believed these were useful, they found them insufficient. Dissection of human cadavers was also necessary. In Alexandria, Egyptian embalming practices may have helped sanction this activity, but not the next step. The Alexandrian physicians agreed with Aristotle that dissection of the dead, while it tells us much about the structure of the body, reveals nothing of living function because there is a fundamental distinction between life and death. One logical conclusion of this reasoning would be to dissect live humans. Celsus pointed out that deliberate vivisection gave more valuable information than chance observation of wounds, which would be in a pathological rather than a normal state. He offered the argument that the sacrifice of a few for the many was justified. In the Rome of Celsus, a criminal could "pay" for his crime by making his body useful to the community.

For empirics and dogmatists, moral arguments based on the undoubted cruelty of vivisection—either of animals or of humans—were much less important than intellectual arguments about the best method of obtaining knowledge about the body. In a later period, early Christian commentators would cite human vivisection in Alexandria as an example of the depravity of pagans. In 280 BCE, human lives—particularly the lives of criminals, slaves, or enemies in war—were not highly valued. Slaves were regularly tortured in courts of law in order to extract evidence, and their mutilation while alive indicated their lower status. Ancient taboos against the mutilation of the body did not take into consideration pain or cruelty.

The Alexandrians derived much of the philosophical basis for their work from the writings and example of Aristotle. Therefore, let us leave Alexandria for a time and travel to Athens, half a century earlier. In 347 BCE, the philosopher Plato died in Athens at the age of 80. It was widely expected that Aristotle, his star pupil, would succeed him as the head of his school, the Academy, but Plato chose a different successor. Aristotle was not an Athenian, and he disagreed with Plato on several fundamental issues. Aristotle left Athens and embarked on a career that changed the course of Western thought.

Plato believed that the external world was but an imperfect imitation of the ideal "forms," or ideas, which could only be perceived intellectually. Aristotle, on the contrary, believed that the form of an object could not be separated from its manifestation in nature. Natural things consisted of both form and matter, a doctrine known as *hylomorphism.* Aristotle thereby established the study of the external world as an important and worthwhile enterprise. Plato's philosophy centered on humans, whereas Aristotle looked at nature as a whole and viewed the human as another animal, although certainly the most highly developed of all. "In all natural things," wrote Aristotle, "there is something wonderful." Even ugly and insignificant ani-

1.1 Human Vivisection

The ancient Roman historian Celsus described the arguments for and against human vivisection. The medieval Wound-man's injuries and weapons were intended to aid the memory of the surgeon. They demonstrate the Empirici's argument that much could be learned in the course of treating injuries. Celsus wrote:

[The dogmatists] hold that Herophilus and Erasistratus did this in the best way by far, when they laid open men whilst alive—criminals received out of prison from the king—and whilst these were still breathing, observed parts which beforehand nature had concealed. . . . For when pain occurs internally, neither is it possible for one to learn what hurts the patient, unless he has acquainted himself with the position of each organ or intestine . . . nor is it, as most people say, cruel that in the execution of criminals, and but a few of them, we should seek remedies for innocent people of all future ages.

On the other hand, those who are called "Empirici," because they have experience . . . contend that inquiry about obscure causes and natural actions is superfluous, because nature is not to be comprehended. . . . But what remains, [it is] cruel as well, to cut into the belly and chest of men whilst still alive, and to impose upon the Art which presides over human safety someone's death, and that too in the most atrocious way. Especially this is true when, of things which are sought for with so much violence, some can be learnt not at all, others can be learnt even without a crime.

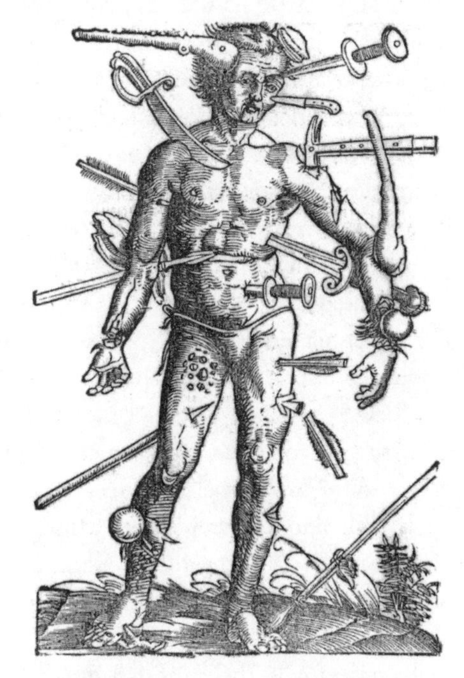

Wound-man, from Hans von Gersdorff, *Feldtbuch der Wundartzney*, 1530. Wellcome Collection

■ Quote from Aulus Cornelius Celsus, *De medicina*, trans. W. G. Spencer, Loeb Classical Library (Cambridge, MA: Harvard University Press, 1960), lines 23–27, 40–43.

mals displayed the creative power of nature, and in his *History of Animals* he attempted to describe every known creature. Most of all, nature revealed its purpose, what Aristotle called the "for something's sake," which led to knowledge of the beautiful, the ultimate goal of Plato's thought.[2] This notion of purpose in nature, known as *teleology* (from *telos*, the Greek word for purpose or meaning in the sense of fulfillment or completion), has proven to be a powerful and enduring idea in Western thought.

To Aristotle, studying animal structure and function was an important way to learn about nature. He could not dissect humans, but he dissected many dead animals, at times killing an animal specifically for that purpose. In addition, he experimented on (and once cut open) live animals, arguing that living function could be explored only in the living. Few others had investigated nature in this way. Animals had been part of religious rituals for millennia, and earlier Greek writers had observed animals and written about them. But Aristotle's interest in their internal structure and function was new.

Aristotle admitted that dissection could be unpleasant. The act of cutting and the look, smell, and feel of the internal parts of the body all were distasteful, and few people were willing to attempt this practice. Dissection, then as now, was also a difficult skill to master, requiring patience, a strong stomach, and manual dexterity. The vascular system, for example, was difficult to discover in dead animals because the vessels collapsed after death. In living animals, the veins usually were not visible to an external observer. The best solution, according to Aristotle, was to starve an animal and then strangle it. Looking at old men had taught him that blood vessels stand out more clearly in an emaciated body, and strangling prevented blood loss.

Following Hippocrates, Aristotle assumed that an analogy existed between human and animal, especially mammalian, structures. Humans were another animal, and knowledge of the human body was Aristotle's goal. The purpose of nature was to fulfill the chain of being, which defined a hierarchical arrangement of nature with humans on top and animals below, descending a ladder, or chain, all the way down to rocks. Each species achieved its proper form and realized its proper potential on a scale. The hierarchy of nature mirrored the hierarchy of society, in which men were superior to women and some humans naturally ruled, while others were natural slaves. Rationality was the key to both hierarchies, for the mind always dominated the body.

But if humans were at the top of the scale because of their superior minds, how could they also be analogous to lower animals? The analogy existed entirely in the body. The nature of the scale gave a clue, for the spaces between species were not large and abrupt, but small and incremental. In many aspects, animal anatomy was like that of humans, as was obvious even from external observation, and animals could legitimately act as stand-ins for examining the inner parts of humans.

A fundamental difference nonetheless separated humans from animals. Aristotle believed that only humans had intelligence and therefore rational souls; animal souls possessed emotion but not reason. Humans and animals therefore did not occupy the same moral plane. He concluded that because animals were not rational and were incapable of deliberate choice, there was no such thing as justice or injustice toward them. Aristotle never spoke about the morality of using animal bodies, either alive or dead, and the use of animals for sacrifice, food, and labor similarly was not a moral issue for him. Such notions would have seemed absurd, as indeed did the notion that all humans, whether male or female, free or slave, citizen or noncitizen, had certain rights.

Theophrastus (ca. 372–287 BCE), Aristotle's collaborator and successor as head of the Lyceum, his school in Athens, took his master's arguments about analogy a step further, arguing that the physical analogy between humans and animals had deeper implications. In contrast to Aristotle, Theophrastus believed that animal sacrifice displeased the gods. He argued that vivisection and meat eating were inhumane: if humans felt pain, so did animals. Humans and animals shared a kinship based on physical and mental similarities; therefore, to deprive an animal of life was an unjust act. Modern ethicists still debate whether loss of life, as well as suffering, is a harm and therefore morally unjust. However, Theophrastus tempered his position in practice with the equally compelling concept of necessity. Sometimes, he said in his treatise *On Piety*, it was necessary to kill animals to eat, and this was permissible. Even if one deplored the spilling of blood, it was sometimes necessary for a society to execute criminals. Only unnecessary cruelty was to be avoided. By this argument, vivisection could fall under the category of necessary cruelty and even human vivisection might also be permissible.

At about the time that Theophrastus wrote, Herophilus and Erasistratus in Alexandria required no philosophical justification for vivisecting humans or animals. Although many later authors condemned them, some, like the Roman physician Galen (ca. 130–210 CE), envied them. In fact, much of what we know of the methods and discoveries of the Alexandrians (whose own works survive only in fragments) comes from Galen's writings.

Galen quoted extensively from Herophilus's treatise *On Anatomy.* Herophilus was the first to distinguish nerves from other tissues, and he went on to establish that the brain rather than the heart (as Aristotle had argued) was the center of the nervous system and that the perception of pain depended on both brain and nerves. His descriptions of the brain—incorporating work on human and ox brains—far surpassed those of Aristotle in accuracy and detail. He also distinguished between sensory and motor nerves. His work on the eye revealed the existence of the optic nerve as well as the retina, which he named. Many of the names Herophilus gave to anatomical features have survived, such as the *styloid process* of the skull, so called for its similarity to a long, slender writing implement known as a stylus. He also offered the first detailed description of the human liver and compared it with the animal liver. The liver was of great importance in ancient Greek medical theory as the supposed source of blood, and in religion, divination by means of the inspection of the liver of a sacrificial animal was a common practice. Moreover, Herophilus clearly described a human liver, not an animal liver as Aristotle had, and his description must have been a result of numerous dissections of humans. Herophilus also offered a detailed account of the vascular system, distinguishing between arteries and veins, which Aristotle had not clearly demarcated.

Herophilus's younger contemporary Erasistratus also worked on the vascular system. His interest was in function as much as in form, and he performed many experiments on live animals, especially pigs, oxen, and goats. He explained the role of valves in the heart and compared the action of the heart to a pump or bellows. These observations were independently rediscovered in the seventeenth century by William Harvey. Erasistratus distinguished three kinds of vessels in the body: veins, arteries, and nerves (which he believed to be hollow). Each of these vessels contained a different fluid or gas: veins contained blood, nerves contained a fluid that conveyed impulses to the brain, and arteries contained air.

How could Erasistratus argue that arteries contained air? Weren't the facts obvious? In dead bodies he saw empty arteries, because in death the blood moves to the veins as it exhausts its supply of oxygen. But in live bodies blood comes from an artery when it is cut. Erasistratus explained that blood rushes into a cut artery from the veins because the air in the artery rushes out, creating a vacuum. He reconciled two apparently contradictory obser-

1.2 The Poison King

Two centuries after the experiments of the Alexandrian physicians, the Greek-Persian monarch Mithridates (134–63 BCE) experimented both on prisoners and on himself. Mithridates discovered that if he took small doses of arsenic, over time he became immune to its effects. He developed antidotes to other poisons, which he tested on prisoners who had been condemned to death. His secret universal antidote became known as mithridatium, a name that lived on although the original recipe did not. Mithridates dosed himself with his secret potion every day, and he became known in history as the Poison King.

More than 1,000 years later, a chronicler accused the Holy Roman emperor Frederick II (1194–1250) of testing poisons on prisoners. Although modern historians have dismissed these accounts as exaggerated, medieval prisoners of war were treated harshly, and the line between torture and experimentation might have been thin.

Mithridatium jars, ca. 1600. Wellcome Collection, CC BY 4.0

vations with a theory that explained both. Later researchers explained this same experimental result differently.

Herophilus and Erasistratus represent the high point of Alexandrian medical research. Human vivisection and dissection in antiquity began and ended with them, and after them animal experimentation also fell into disuse for a time. The empirics won the battle of the medical sects, and both Greek and Egyptian taboos against mutilation of the body reasserted themselves. Later kings in Alexandria were less receptive to Greek scientific ideas. In premodern societies, the cultural sanction of scientific activity was often fragile and fleeting. Galen alone, some 400 years later, reached the Alexandrian physicians' level of expertise, and many of their findings were not confirmed until the seventeenth century, when conditions for research were comparable.

Galen, Gladiators, and Apes

Galen, physician to Emperor Marcus Aurelius, had declared in his tract that "blood [not air] is contained in the arteries" (*an in arteriis natura sanguis contineatur*). A follower of the ideas of Erasistratus issued a public challenge and attempted several public demonstrations to prove his point; Galen in turn accepted the challenge, experimenting on many live animals. In the most telling experiment, Galen and his students cut open an animal—probably a pig—to expose the aorta, tied off the artery in two places, and after the death of the animal cut the artery between the ligatures. Blood, not air, poured out. The audience no doubt applauded.

There was indeed an audience, for neither Galen nor his challengers were about to resolve a public challenge in private, and anatomy was one of many forms of entertainment in the crowded streets of ancient Rome. Galen had already publicly dissected an elephant, not once but twice. In the highly competitive arena of Roman medicine, reputations could be made or broken with the stroke of a knife. (Galen had left Rome under a cloud in 166 CE, only to return triumphantly with the emperor three years later.) The rivalry of public dissection was the medical equivalent of gladiatorial combat, a practice with which Galen was quite familiar, since he had spent the early years of his medical career as a physician and surgeon at the school for gladiators in his native town of Pergamum in Asia Minor.

Galen was, with Hippocrates, one of the two most important physicians in antiquity, and his ideas dominated Western thought until the Renaissance. Along with the Alexandrians, he was also one of the greatest anatomists and experimenters of his era. The rediscovery in the sixteenth century of his manual for dissection, *On Anatomical Procedures*, introduced his methods of dissection and animal experimentation to a new generation of researchers and established these methods as the only legitimate means of obtaining knowledge of animal and human bodies. Without Galen, Harvey could not have made his discoveries.

Galen spent much of his early life as a peripatetic student of medicine, visiting many sites in the empire, including Alexandria. Human dissection had long ceased to be performed there, but Alexandria's medical school remained the most famous in the Western world. Along the way, Galen mastered Greek philosophy and attained impressive skills in anatomy. He perfected the latter as the gladiators' surgeon, learning anatomy, as the empirics advised, by observing wounds.

But in his many books Galen severely criticized the empirics, aligning himself instead with the dogmatic school. The physician, he wrote in his *On Anatomical Procedures*, was not a mere craftsman but a man of learning. Knowledge of inner structures and functions was essential to successful medical practice, and direct experience was the best teacher. Yet Galen did not disdain book learning, and he was as much a synthesizer as an original researcher; often the only surviving evidence of the writings of earlier practitioners is to be found in Galen's works, which fill twenty hefty modern volumes, and that is not his entire production. The works most concerned with the topic of animal and human experimentation are *On Anatomical Procedures*, which contains the text of lectures on anatomy that he delivered in the year 177 CE, and *On the Usefulness of the Parts of the Body*, a general description of the human body written about 175 CE. He wrote all his works in Greek.

Galen believed that all living beings have *physis*, or life. In addition, animals possess what he called *psyche*, or consciousness. Humans possess both these qualities, but they also have the capacity to reason. Like Aristotle, Galen sought evidence of purpose in nature. He argued that each part of the body has a purpose—its "usefulness"—and that it was the job of the anatomist to determine that purpose. The body as a whole fit into the overall plan of the great chain of being. In *On the Usefulness of the Parts of the Body*, Galen aimed to provide a complete description of the human body, and the work is a remarkable performance. However, he neither dissected a human nor witnessed such an event. While he made use of the work of his predecessors, especially the Alexandrians, he based many of his conclusions on animal dissection and vivisection.

In Galen's Rome, human dissection was forbidden. The naturalist Pliny the Elder (23–79 CE) wrote disapprovingly of those who employed human body parts for folk cures, such as scraping sore gums with the tooth of a man who had died violently. In his writings, Galen wistfully mentioned those long-ago days in Alexandria when human bodies could be opened to the physician's gaze. In his day, the best he could do was to dissect a Barbary ape, perhaps with a naked slave nearby to serve as a comparison. The analogy between humans and animals, based on anatomical similarities, was of course a central tenet of the great chain of being and provided another guarantee of the regularity and planning evident in the universe. Galen's extensive use of animals reinforced and strengthened the analogy between animals and humans, making a virtue of necessity and firmly establishing animal experimentation as the standard method for learning about human anatomy and physiology.

In *On the Usefulness of the Parts of the Body*, Galen offered a complete system organized around the central philosophical principle of teleology: the "usefulness" of the parts comprised both their function and their purpose, following Aristotle. Both this principle and his reliance on animal anatomy led Galen to some erroneous conclusions about human anatomy, which critics in the Renaissance later pointed out.

At this point in history, medical theory, in its symmetry and correspondences, demonstrated the meaningfulness of nature as a whole. It was based on Aristotle's concept of four elements (fire, air, water, and earth) and four qualities (hot, wet, cold, and dry). In Galenic medical theory, four humors—phlegm; blood; black bile; and yellow bile, or choler—governed health by their surfeit or lack. Each humor derived from a particular organ—the brain, the heart, the spleen, and the gall bladder, respectively—and governed a particular temperament. The words *phlegmatic*, *sanguine*, *melancholy*, and *choleric* still describe certain mental states, even though we no longer believe in the theory of the humors. When a person was in perfect health, the humors existed in equitable quantities and in harmony with one another, but an imbalance would cause illness. Because each person's natural balance and temperament were unique, each experience of illness was individual. Generalizations about disease, which the empirics hoped to acquire, were therefore of limited value. Repeated encounters with patients, such as those described in the *Epidemics* of Hippocrates, could help the physician to identify salient signs, but treatment had to be carefully tailored to the individual. In addition, many other factors, including weather, geography, and astronomical phenomena, might contribute to an illness.

Therapies, therefore, concentrated on restoring balance, often by administering medicines or treatments opposing the effect of the excessive humor. Thus colds, caused obviously by excessive phlegm (a cold, moist humor), responded to hot, dry medicines. One sign of fever was a flushed face, caused by an excess of hot, wet blood. A common therapy was to let blood out of the patient, restoring the proper balance, and the patient did become cooler and pale as a result. Regulated diets and bathing (hot or cold) were also important therapies.

The physiology that Galen detailed in *On the Usefulness of the Parts of the Body* supported this theory of illness and therapy. He viewed digestion as a critical function that helped to establish the balance of the humors; too much of the wrong kind of food could cause all manner of ill health. His description of the process by which digested food became blood emphasized the separation from the blood of the two forms of bile and their proper placement in the spleen and the gall bladder. He also spent much effort on describing the process by which wastes were eliminated from the body, another important aspect of therapy. In addition, his detailed anatomical descriptions of limbs and of bone, skin, and tendon had direct application to the treatment of wounds, and Galen spoke harshly to those who would ignore dissection and needlessly injure their patients out of ignorance.

Nonetheless, much of Galen's work in anatomy had little direct application to medical theory or practice, and when he listed the uses of anatomy, medical practice came last. He constantly searched for purpose in nature, and he dissected many different kinds of animals, he said, to show that a single mind had conceived them. Equally important was what Galen called "knowledge for its own sake."[3] Although he declared near the beginning of *On Anatomical Procedures* that the most important function of anatomy was to enable the physician to treat wounds, this theme quickly receded before the thrill of discovery.

As Aristotle had discovered, dissection was not easy, and in the meticulous descriptions in *On Anatomical Procedures* we see a master at work. No detail was too small or unimportant for Galen; he advised the diligent dissector to remove the skin of the animal himself, for careless assistants could damage important structures. This attentiveness in dissection trained the student for the more difficult task of vivisection. Galen's accounts of vivisection do not make for easy reading. His matter-of-factness, his brash delight in besting his intellectual enemies, and the clinical detail of his accounts still disturb readers two millennia later. Nevertheless, in Galen's work we can see the beginnings of the Western tradition of biological research, and his own utter absorption in his task draws the reader along, however unwillingly.

Although therapeutic benefit received an occasional mention—and the research described in *On Anatomical Procedures* formed the basis for his therapeutics in *On the Usefulness of the Parts of the Body*—there is no mistaking Galen's passion for research as an end in itself. When he described opening the chest of a living animal to view its beating heart, the hemorrhage from cut arteries was simply an annoyance; when the thorax was cut in a certain way, he noted, the animal ceased to breathe or cry out, but if the anatomist covered the cuts with a hand, breathing and crying out resumed. If certain nerves were cut or damaged, the animal would lose its voice entirely, or if merely half cut, the animal would lose only some of its voice. His interest was in the effects, not in the animal, which was just a tool, a means to an end.

The nervous system, and the effects of cutting certain nerves, fascinated Galen. Severing the spinal cord in pigs, goats, and apes, he noted the degree of paralysis produced. He also showed which functions were controlled by which nerves. He then went on to study brain function: which part of the brain controlled which part of the body? Could he convince those who doubted that the brain, and not the heart, controlled the nervous system? Dissecting ox brains given to him by a butcher was practice for the main event: the vivisection of the brain of an ape, in which he stimulated various parts of the living brain and noted the effects produced on the body.

Galen advised his students to cut "without pity or compassion" into a living animal. He explained that the cries of an animal in pain were part of the procedure, and the "unpleasing expression of the ape when it is being vivisected" was unavoidable.[4] Galen had little concern for the animals, and by modern standards he was very cruel. His world was one of cruelty and violence in which gladiatorial combat was as common as football, animals were tormented and killed in arenas for amusement, and one form of public execution was to throw condemned humans together with wild animals into the amphitheater for an afternoon's entertainment.

Because animals ranked below humans in the great chain of being, Galen followed Aristotle in granting to animals only limited consciousness, which also implied considerably less consciousness of pain. An animal whose thorax had been opened, said Galen, was perfectly unimpaired in its functions after the operation ended. He compared its state to that of a slave who had lost part of his sternum yet survived: "It is surely more likely," wrote Galen, "that a non-rational brute, being less sensitive than a human being, will suffer nothing from such a wound."[5]

Galen discovered more about the animal body, and by analogy the human body, than had ever been known before. He was the last great orig-

Title page from Galen, *Opera omnia*, 1565. Wellcome Collection

1.3 Dissecting a Pig

Galen dissects a pig in this illustration from a sixteenth-century edition of his works. He described his method of experimentation in this passage from *On the Natural Faculties*:

The fact is that those who are enslaved to their sects are not merely devoid of all sound knowledge, but they will not even stop to learn! Instead of listening, as they ought, to the reason why liquid can enter the bladder through the ureters, but is unable to go back again the same way—instead of admiring Nature's artistic skill—they refuse to learn; they even go so far as to scoff, and maintain that the kidneys, as well as many other things, have been made by Nature *for no purpose*! . . . We were, therefore, further compelled to show them in a still living animal the urine plainly running out through the ureters into the bladder; even thus we hardly hoped to check their nonsensical talk.

Now the method of demonstration is as follows. [Galen then described his procedure on a living animal, tying off the ureters to demonstrate that they pass urine from the kidneys to the bladder. The ureters were then untied and the bladder was allowed to fill, and the penis was tied off to prevent urination. The urine, however, remained in the bladder, even when it was squeezed, and did not return to the kidneys.] Now, if anyone will test this for himself on an animal, I think he will strongly condemn Asclepiades [Galen's rival], and if he also learns the reason why nothing regurgitates from the bladder into the ureters, I think he will be persuaded by this also of the forethought and art shown by Nature in relation to animals.

■ Quote from Galen, *On the Natural Faculties*, 1.13, trans. A. J. Brock, in *Source Book of Greek Science*, ed. M. R. Cohen and I. E. Drabkin (Cambridge, MA: Harvard University Press, 1948), 481–82.

inal thinker in the Greek tradition, and animal experimentation in his manner was not performed in the West for over a thousand years after his death. By chance and circumstance, his works survived when the works of others did not, and he became a major authority in the Christian and Islamic worlds when his works were translated from their original Greek into Latin and Arabic. He established standards, techniques, and an attitude toward animal experimentation that in many ways still inform the science of today. Galen emerges from his works as a recognizable personality, if not necessarily an attractive one. He forces us to confront the consequences, both good and bad, of animal research.

The Christian Body: Angel or Brute?

Galen seems an unlikely candidate for Christian embrace. He was not a Christian, and his boastfulness and cruelty have little connection with the biblical "gentle Jesus, meek and mild." Yet Galen's work was adopted by medieval and Renaissance Christians for several reasons. He was deeply religious and celebrated in his work the creativity and miraculous organization of nature, which he attributed to the guidance of an active intelligence, the Platonic creator-god. The last book of *On the Usefulness of the Parts of the Body* is a paean to the skill of this intelligence, whom he referred to as "Creator" or "Nature" but who could very easily have been viewed as God the Creator of Judeo-Christian scripture. In addition, *On Anatomical Procedures,* the work most revealing of Galen's personality and most explicit regarding his treatment of animals, was not translated from Greek into Latin until the 1530s and was therefore inaccessible to most Western readers after the fall of Rome in the fifth century.

Were Galen's views on animals incompatible with Christian doctrine? What did Christian doctrine have to say about animals? The Christian's relationship to the natural world has been debated since the foundation of Christianity, with implications for modern environmentalism as well as animal welfare. The Bible is by no means straightforward on this issue. In the book of Genesis, God commanded the first humans to "be fruitful and multiply, and fill the earth and subdue it; and have dominion over the fish of the sea and over the birds of the air and over every living thing that moves upon the earth." Later in Genesis, Adam named the animals, conferring upon them identity in terms of his dominance.[6] However, that dominance was not initially one of plunder or indiscriminate consumption. Before the

Fall, Adam and Eve were vegetarians, and after, God valued the farmer Abel more than the herdsman Cain.

The Hebrew Old Testament includes several passages that contradict the edict to dominate nature. For example, passages in Psalms express an appreciation of, rather than a dominance over, nature. The writer of the book of Ecclesiastes declared, "For the fate of humans and animals is the same; as one dies, so dies the other. They all have the same breath, and humans have no advantage over the animals."[7] Some early Jewish scholars interpreted such passages to mean that humans were stewards of the earth, entrusted with its care and preservation by God, who alone possessed and ruled it. Because God had created humans to complete his creation, each generation should leave the earth more beautiful and fruitful than before. This is an agrarian ideal of a domesticated nature, encompassing Abel the farmer and Cain the herdsman. This ideal nature is tame, not wild.

A more prominent view, held by many early Christian theologians, argued that because God had made humans in his image, unique in their possession of an immortal soul, they were naturally above the rest of nature. This view was compatible with ancient ideas such as the great chain of being, which also placed man at the top of creation. The Greek philosophical school known as the Stoa similarly argued that rationality distinguished humans from the rest of nature and made them closer to the gods. Like the gods, humans therefore enjoyed a natural hegemony over nature, which was made for their use.

Christianity drew both from the Hebrew tradition and from other ancient religious and philosophical traditions. Central to Christian doctrine was its recognition of the value of the individual, whose salvation was the result of his or her behavior in this life. Nevertheless, there was little consensus on what constituted Christian behavior toward animals and the natural world. The New Testament, like the Old, gave no clear guidelines. Jesus employed animals as symbols of faith—the hen gathering her chicks under her wing, the shepherd who rejoices in finding the stray sheep—but he assured his followers that their lives were worth more than those of the beasts.[8] He transferred demons from humans to pigs and allowed these Gadarene swine to run to their death. The evangelist Paul also stated in his first letter to the Corinthians that God's main concern was with humans, not animals. The ultimate concern of Christians was to achieve life after death, and this life was accessible only to humans, who possessed immortal souls.

Many early Christian commentaries emphasized God as creator and discerned the nature and purpose of God insofar as they detected purpose and meaning in nature, in a manner similar to Galen's in *On the Usefulness of the Parts of the Body.* The North African bishop St. Augustine (354–430), the most influential of the early church fathers, developed this theme. In his *City of God* he declared that God "did not wish the rational being, made in his own image, to have dominion over any but irrational creatures, not man over man, but man over the beasts."[9] Augustine's views on nature are part of his larger picture of fallen humanity, for nature reflected the consequences of Adam's fall, which changed a perfect landscape into one linked inextricably with death and suffering. Although nature continued to demonstrate God's design and purposes, this design had been corrupted by human sin. The pervasiveness of sin gives a sinister cast to Augustine's introduction of the concept of the "two books," according to which the "book of nature" complemented the book of God, the Bible. This concept was later used in Christian Europe to defend the practice of science.

Augustine acknowledged the chain of being in his definition of human dominion. Humans, he said, fall between animals and God on the scale, but reason makes them closer to God. While animals are superior to plants, humans are superior to all. Each being on the scale has value as a creation of God, but all are not of equal value. To think that all were of equal value was "the height of superstition," according to Augustine. "There are no common rights between us and the beasts and trees," he wrote, adding that "we can perceive by their cries that animals die in pain, although we make little of this since the beast, lacking a rational soul, is not related to us by a common nature."[10] This remained the Christian doctrine for the treatment of animals until modern times.

Two Christians of the Middle Ages, St. Francis of Assisi and St. Thomas Aquinas, revisited Augustine's doctrines in ways that would be important for the experimenters of the sixteenth and seventeenth centuries, who revived the techniques of Galen and the Alexandrians. In the same period, St. Albert the Great opened the way for the Christian study of Aristotle's works, including his work on animals. St. Francis of Assisi (1182–1226) has been praised as the first environmentalist. However, a closer look at his work reveals that he viewed animals as part of a human-centered symbolic system. The *Fioretti*, a popular book of stories about Francis, included many instances of his kindness to animals. He preached to the birds, he rescued doves from a hunter and built them nests, and he persuaded the fierce wolf of Gubbio to stop attacking the townspeople. In "The Canticle of Brother Sun," he praised Brother Sun and Sister Moon. In all these stories, the larger context is clear. The birds to whom Francis preached demonstrated the creativity of God and symbolized the Franciscan friars, who owned nothing and traveled throughout the world. The wolf of Gubbio served as a reminder to the sinful townspeople of the perils of hell. The doves represented pure souls whom Francis rescued from sin. Brother Sun is a symbol of God himself. An example of the everyday treatment of animals is the story of Brother Juniper, who cut off the foot of a living pig to feed a sick friar. Francis scolded Juniper, not for his cruelty to the pig but because the pig belonged to someone else.

Francis lived during the High Middle Ages, when Europe gained peace and prosperity after the centuries of turmoil that followed the fall of Rome. He could wander around begging without worrying about bandits; he could walk in the woods and preach to the birds without fear. The wolf of Gubbio was a lingering vestige of the old fear of the wilderness, the wild animal leaping into your living room. But as it turned out, he was not very formidable, and like the rest of nature, he could be tamed by Christianity and civilization.

During the lifetime of Francis, an immense project of translation and transmission was under way that reintroduced the works of Aristotle, along with some works of Galen, to western Europe. The impact of Aristotle's philosophy of nature on medieval intellectuals was immediate and overwhelming. The interests of St. Albert the Great (known as Albertus Magnus, ca. 1200–1280) ranged from theology to natural science, and he worked to reconcile Aristotle's ideas about nature with Christian theology in treatises on minerals, plants, and animals. Albert's *De animalibus* (On animals) combined commentaries on Aristotle's books on animals with an alphabetical encyclopedia of animals, including his own firsthand observations.

Albert's most famous pupil, the eminent theologian and scholar St. Thomas Aquinas (1225–1274), delineated the realms of faith and reason in his *Summa theologica.* Because reason was God given, it could lead the faithful to truths compatible with God's truth of revelation. He reiterated Augustine's "two books" notion, agreeing that the book of nature complemented the book of God. Aristotle's teleology supported Thomas's theology: nature was created for a purpose, and that purpose was good. Therefore, nature was good, and its study beneficial. In this period, anatomical studies began to resume in some Italian schools of medicine, and dead animals were the anatomical subjects.

Thomas, like Augustine, also accepted Aristotle's hierarchical view of nature as a chain of being. In the *Summa theologica*, Thomas explained that human superiority stemmed from the possession of reason, which implied the existence of an immortal soul. Since animals lacked reason, they lacked immortal souls and could not participate in the afterlife. Because they were irrational, they also lacked the ability to choose. Animals appeared to make choices because of the structure of their parts, not because they exercised reason and free will. Thomas compared animals to the new mechanical clocks of the time. If humans could craft such devices, how much more sophisticated were the machines of God? The machine analogy would prove to be long-lived.

Thomas believed that animals were created for the sake of humans. Reason tells us, he said, that there are no restrictions on the human use of creatures. On earth, all things exist for human use, to help them to become closer to God. But cruelty can have no role in the godly character. Therefore, although there was no theological stricture against it, good Christians, who were compassionate, would treat animals well. This action did not in itself make humans good, because animals lacked moral status. Goodness was defined solely in terms of behavior toward other humans. Nonetheless, argued Thomas, cruelty to animals could lead to cruelty toward humans, although the definition of what constituted cruel behavior varied widely. Thomas's argument became a standard reference for the treatment of animals and established a theological basis for experimentation on them, although this was not practiced in his time. When experimentation revived during the Renaissance, justification was readily at hand.

In the ancient and medieval periods in Western culture, Greek concepts about the body and nature joined together with Christian notions about the place of humans in nature and their duties toward the natural world. Together, this powerful combination of ideas formed the framework of the Western point of view toward animals, humans, and experimentation. Although particular ideas were later disputed, the general framework, which declared the superior status of humans among living beings, remained and influences our ideas even today.

2 Animals, Machines, and Morals

Imagine a chilly winter day early in 1621. The doctor strode into the room and flung back the shutters, letting in the thin sunlight. The room was cold, but he did not notice it as he moved rapidly about, setting out tools and instruments. He walked over to a stack of cages and opened the topmost one, gently pulling out a large rabbit. Its nose twitched as the doctor carefully tied it to a board with holes drilled in it through which he passed the thin cord that bound the animal's limbs. The rabbit lay on its back, blinking and quivering, its limbs splayed, its chest rising and falling quickly. The doctor took a sharp, thin-bladed knife and with practiced skill laid open the rabbit's chest. The animal struggled and panted, but the bonds held fast. The doctor sliced through the breastbone and spread open the ribcage with his strong fingers, exposing the rapidly beating heart. He cut a strong silk thread and tied it around the rabbit's aorta, watching with satisfaction as the animal's heart grew engorged with blood, while the vessel beyond the ligature became white. He delicately sliced the aorta and saw the blood spurt out in regular pulses. As the rabbit slowly expired, the doctor seized a notebook and began to write, looking up to observe the rabbit's heart as it slowed.

William Harvey (1578–1657) repeated this and other experiments hundreds of times, on dozens of different animals, to prove his theory that the blood circulated through the body. Harvey was the first since Galen to initiate a research program based on experimentation on live animals. His discovery of the circulation of the blood is the most important event in the history of medicine and marks the beginning of medicine as a science. His research methods combined animal dissection and vivisection with experiments and quantitative arguments and modeled future research. During the seventeenth century, researchers across Europe employed Harvey's methods. Unlike Harvey, these experimenters believed that animal and human bodies were machines that could be analyzed in mechanical terms. They called themselves natural philosophers and did not fully distinguish the biological from the physical (the term *biology* dates from the nineteenth century). The many versions of this *mechanical philosophy* significantly increased the understanding of the body and its functions. Parallel to this mechanization of the body, and partly in reaction to it, a new sensitivity toward and awareness of animals also began to evolve in this period.

An experiment involves foresight, planning, and the prediction of an outcome (the hypothesis); it entails active intervention in natural processes, not merely passive observation. When Harvey tied a ligature around the rabbit's aorta, he controlled the conditions of the experiment by actively intervening in the normal processes of nature. His hypothesis, that the blood flowed in one direction, guided this intervention; blocking the flow led to certain consequences. An experiment should be repeatable, with the same results each time, before it can be accepted as a demonstration of how nature works. Harvey repeated his animal experiments many times over several years before he was satisfied that what he had found was indeed a fact about nature. He therefore assumed that nature followed certain laws and that a consistent effect implied a consistent cause. Nature was not arbitrary in its actions. As a Christian, Harvey probably believed that miracles could occur but that they were outside the ordinary course of nature, which in its lawfulness displayed the intelligence of its creator.

By proposing that the blood circulates throughout the body, Harvey advanced a theory that completely restructured his field of inquiry. He presented a new model for science that entirely changed the rules for the conduct of research. Galen believed that the blood did not circulate, but moved back and forth in the blood vessels, and that the liver constantly made new blood to replace what the body's organs used up. This concept of the vascular system dictated the questions other researchers could ask about the blood. According to this explanation, venous blood was new blood as it came from the liver, dark in color and full of nutrients. The lungs served only to cool the warm heart. The Galenic model accommodated new evidence discovered in the sixteenth century by anatomists in Padua, where Harvey studied. In 1543, Andreas Vesalius (1514–1564) had demonstrated that the septum, the muscle dividing the left and right sides of the heart, was solid, not perforated as Galen had claimed. However, he did not go on to question how the blood then got from the right side of the heart to the left. A decade later, Realdo Colombo (ca. 1515–1559) concluded from dissection and vivisection that the blood passed from the right side of the heart through the lungs to the left side, what we now call the pulmonary circulation. Nonetheless, Colombo continued to believe that the lungs functioned primarily to cool the blood.

To Harvey, the pulmonary circulation was an anomaly that did not fit Galen's explanation. Rather than trying to fit new evidence into an old theory, Harvey decided that the old theory was incorrect and that a new explanation of the observed phenomena was necessary. His experiments helped him to devise a new theory—that the blood circulated and that it was reused rather than constantly newly made—and to test it. Until Harvey challenged the prevailing model, no one performed these tests, such as the ligature experiment described at the beginning of the chapter, because they did not answer questions that were meaningful to Galen's model.

Harvey used other phenomena that had not previously been viewed as anomalies as further proof of his new theory. One phenomenon was the existence of valves in the veins. Harvey's teacher Fabricius (Girolamo Fabrici d'Acquapendente, 1537–1619) thought that these valves simply slowed the back-and-forth flow of blood to extract its maximum nutritive value. Harvey, however, believed that the valves kept the blood flowing in one direction, as his theory of circulation required. His experiments followed a modern pattern: one experiment led to another, in a chain or tree. In addition, Harvey developed multiple tests to demonstrate a single experimental fact. He demonstrated the purpose of the valves in the veins not only by inserting a probe into the vein of a dead animal—which showed that the valves only opened one way—but also by externally observing the flow of blood in the arm of a human subject with especially prominent veins.

The University of Padua in northern Italy, where Harvey studied at the end of the sixteenth century, was the best place in Europe for anatomical study. During the fourteenth and fifteenth centuries, human anatomical demonstration for instruction had become commonplace in universities across Europe. These formal demonstrations included the dissection of dead animals. Medieval natural philosophy emphasized the authority of the text as the source of knowledge, and anatomical demonstration long served to illustrate the text—some version of Galen—rather than to yield new information. The rediscovery of additional ancient texts in the fifteenth and sixteenth centuries, however, eventually led to major changes in ideas about the body and how knowledge is gained.

Medieval scholars discovered classical texts through a complex process of retrieval and translation, often from Arabic sources. Renaissance scholars sought out original Latin and Greek texts and found many previously undiscovered ones. For example, the rediscovery of the medical history of Celsus in 1478 revealed the many conflicting schools of ancient medical thought, and the rediscovery of Galen's *On Anatomical Procedures* in 1531 prompted a revival of animal experimentation. In addition, Renaissance artists' passion for realism led many of them to attend anatomical demonstrations and even to dissect animals and humans themselves. In his unpublished notebooks, Leonardo da Vinci (1452–1519) described dissections and vivisections that he witnessed and performed.

In 1543, Vesalius published his response to Galen, *On the Fabric of the Human Body in Seven Books* (usually referred to as *De fabrica*, from its Latin title, *De humani corporis fabrica humani libri septem*). Appointed professor of anatomy at Padua at the age of 23, Vesalius dissected and vivisected for research as well as for classroom demonstration. University anatomy lecturers had use of the bodies of a certain number of executed criminals for their annual public lectures. But Vesalius required many more bodies for his research program. At this time, the practice of human dissection was restricted by physical, political, and social circumstances. The difficulty of preserving bodies dictated that dissections take place in winter. Local governments granted the use of executed criminals for public dissection but did not provide additional bodies for research purposes. Even though there were no laws or religious rules against human dissection, most people remained reluctant to volunteer their own bodies or those of their loved ones to be mutilated after death. Although professors of anatomy performed autopsies to determine the cause of death, these did not allow for extensive research.

Vesalius frequented executions in the hope of obtaining a body fresh from the hangman. His students held "resurrection parties" to rob new graves. Nevertheless, the supply of bodies, particularly of female bodies (since most executed criminals were males), continued to be inadequate for his purposes. For example, in his research on the female reproductive system between 1537 and 1543, Vesalius had access to only six female bodies. Three of these

Title page from Vesalius, *De fabrica humani corporis*, 1543. National Library of Medicine

2.1 Showman before a Raucous Crowd

The title page of Vesalius's landmark work *De fabrica* (1543) depicts him at center stage, his hands poised in mid-dissection. Like Galen, Vesalius is a showman before a raucous crowd. Unlike in the past, no professor hovers above the audience reading a text, no assistant does the manual work of cutting. Vesalius was both lecturer and dissector and the equal of the ancients, represented by classically garbed observers. A dog and a monkey await dissection, or vivisection, in their turn; in a 1540 demonstration, Vesalius used three human bodies, six dogs, and other animals as well.

were for demonstrations and therefore were limited in their usefulness for research; one was the body of a child stolen from its grave; one was the body of a murdered woman on which Vesalius had performed the postmortem. Therefore, only one body was available solely for research. Although Vesalius criticized Galen's reliance on animals to explain human anatomy, it is not surprising that Vesalius too relied heavily on animals as stand-ins for humans. He refuted many of Galen's claims that had been derived from animal anatomy, but the lack of correspondence between animal and human anatomy on some points did not discredit other claims. Vesalius devoted the last chapter of *De fabrica* to the dissection of living animals, agreeing with Galen that function, as opposed to form, could not be fully understood without recourse to living animals.

Vesalius's boastful accounts in *De fabrica* of his methods of obtaining bodies led to the enactment of laws against grave robbing and to fears on the part of many Europeans of disrespectful mutilation after death and even accidental (or not so accidental) vivisection. Vesalius omitted some of the more offensive passages from later editions of *De fabrica*, but the damage had been done. In addition, the naturalistically posed, anatomized humans in Vesalius's illustrations presented the human body, much like that of any other animal, as an object for the gaze and knife of the anatomist, blurring the boundary between the animal and the human body.

When Harvey traveled to Padua in 1599, Fabricius held the anatomy chair once occupied by Vesalius. A devoted Aristotelian, Fabricius viewed the animal body as a universal concept built from many observations of individual animals, including humans. His work on vision, voice, and hearing (*De visione, voce, et auditu*, 1600) compared several different animals. Aristotle's search for cause and purpose also motivated both Fabricius and his pupil Harvey.

Harvey pursued his anatomical researches privately after he returned to London in 1602. Circumstances dictated his use of animals, since human cadavers remained scarce. Philosophically, following Fabricius, Harvey aimed to understand animals in general, not only humans. He set out to learn about the motion of the heart, and the idea of circulation gradually emerged. Harvey presented some of his conclusions in public anatomy lectures delivered to the London College of Physicians (the physicians' guild, not an educational institution) from 1616 to 1618. Like most anatomy lectures, these included both the dissection of a human cadaver and the dissection and vivisection of several animals. His lectures were well attended, and a critic complained that Harvey did not restrict his audience to the learned: "If only, Harvey, you would not hold an anatomy in front of jacks-in-office, petty lordlings, money-lenders, barbers and such like ignorant rabble, who, standing around open-mouthed, blab that they have seen miracles."[1] Harvey gave no hint of circulation in these lectures. He continued to experiment, to make and discard hypotheses, and by the early 1620s a theory of circulation had begun to take shape.

Harvey revealed this theory in *Anatomy of the Motion of the Heart and Blood in Animals* (known as *De motu cordis*, from its Latin title, *Exercitatio anatomica de motu cordis et sanguinis in animalibus*), published in 1628. Here he proved three hypotheses. The first is a quantitative argument: the amount of blood that passes through the arteries is much greater than the body could produce from a normal amount of nourishment. Harvey computed the amount of blood driven through the heart, using anatomical evidence for the capacity of the heart. Each beat of the heart, he estimated, drove out half an ounce of blood; at 2,000 heartbeats an hour, this added up to more than 60 pounds an hour, or nearly 1,500 pounds a day. In animals this discrepancy was even more striking, for Harvey calculated that 3.5 pounds of blood passed through the heart of a sheep in half an hour; and as he found by bleeding a sheep, its entire quantity of blood was only 4 pounds. The body could not consume and constantly make so much blood; therefore, it must have been reused.

Harvey's second hypothesis was that the blood is driven to all parts of the body through the arteries by the action of the heart. He defined systole, or contraction, as the active phase of the heartbeat, in contrast to ancient writers, who believed that expansion, or diastole, was active. At diastole, Harvey argued, the heart was at rest. The cause of the pulse was the expansion of the arteries as the contracting heart filled them with blood. Harvey's third hypothesis, that the blood is returned to the heart by the veins in an equal amount to that pumped out to the arteries, logically followed, although he could not observe the minuscule connections between the veins and arteries.

Since Harvey sought general principles applicable to all animals, he dissected all sorts of hearts, with two, three, and four chambers, belonging to cold- and warm-blooded animals. The hearts shared certain characteristics. For example, all possessed valves that acted as one-way doors to prevent the backward flow of blood. The valves in the veins operated in the same way. Harvey concluded in a very Aristotelian way that structures such as the valves must have some active physiological purpose.

As we have seen, Harvey tied off veins and arteries to observe the surrounding changes. The left side of the heart became distended when the aorta, the outgoing artery, was tied off. Tying off the vena cava, which sent venous blood into the right side of the heart, caused the vein to become distended and the heart to turn whitish from lack of blood. Harvey also noted that a human arm became cold and blue when he tied a tight tourniquet that blocked the arterial flow. Over and over, Harvey cut an artery near the heart of a living animal and saw the blood forcibly pushed out at each heartbeat. The heart, he announced, acted like a pump.

Harvey employed both ancient and modern methods. He performed experiments to discover new facts about nature and to demonstrate facts that he believed he had already proven or that were generally known. Like Aristotle, Harvey also employed many analogies, reasoning that events or structures similar in some ways would be similar in other ways as well. Aristotle and Harvey believed that a fundamental analogy existed between human and animal structures, so that experimenting on animals could yield information about the human body.

Harvey drew another analogy between the macrocosm (big world), or the structure of the heavens, and the microcosm (small world), or the structure of the earth, and between both and the smaller microcosm of the human and animal body. He believed that the circulation of the blood was analogous to larger cycles in nature, including the motion of the heavens and the cycle of rain on earth. The circle is a recurring image in his work. To the Greeks it represented perfection. The heart acted like the sun, which caused the cycle of rain. The sun's warmth caused rejuvenation and growth. Harvey believed that just as the heart and the sun acted as monarchs of their realms, God had given earthly kings the right to rule. Circulation of the blood and the cycle of rain also emulated the perfect, eternal circular motion of the heavenly bodies. By means of these analogies, circulation displayed the meaning and purpose of the universe.

Harvey's significance for science lay not only in his discovery of circulation. He revealed the immense power of experimental demonstration, especially animal experimentation, to disclose valuable information about the body. In his second major work, *Exercises on the Generation of Animals*

(*Exercitationes de generatione animalium*, 1651), Harvey employed his experimental and observational skills to uncover the secrets of generation, including the relative roles of male and female, how fertilization occurs, and the early development of the embryo. He refuted the "erroneous and hasty conclusions" of the ancients; "like phantoms of darkness," he claimed, "they suddenly vanish before the light of anatomical inquiry."[2] Harvey investigated frogs, snakes, fish and other marine animals, insects, and a number of domestic fowl and mammals. Like many others, he traced the development of the embryo by examining the stages of development in chicken eggs. Harvey concluded that the egg was the unit of development for all animals. The title page of *On Generation* proclaimed that "ex ovo omnia" (all is from the egg).

As a physician to the English king Charles I, Harvey had access to almost unlimited numbers of deer for dissection. An avid hunter, the king shot deer at one of his many estates nearly every week. Harvey's work on these animals, coupled with vivisections of dogs, cats, and rabbits, allowed him to give a minute account of fertilization, gestation, and embryonic development. From his observations of one stage of development Harvey devised questions and hypotheses that he then attempted to demonstrate on animals. This close observation of large numbers of animals was as important to subsequent research as were the experimental techniques of *De motu cordis.*

Harvey displays none of the showman's glee of Vesalius in his descriptions of vivisection. For Harvey, animals served a purely instrumental function, and if the question of cruelty occurred to him, he never expressed it. A modest and appealing persona emerges in his books, but Harvey did not hesitate to dissect the body of his own father (in front of an audience) in his quest for knowledge. Such an act was more shocking to contemporaries than the vivisection of animals. In this era, public dissection and vivisection flourished alongside many forms of public entertainment based on animal performance. Dancing bears, bears fighting with dogs, cockfighting, and all manner of cat torture were commonplace, and everyday cruelty to animals was the rule rather than the exception. Petkeeping was still uncommon, although Harvey described his wife's pet parrot, whose death she grieved. Her feeling for the bird, however, did not prevent her spouse from dissecting it.

Like Aristotle, Harvey was a vitalist who believed that living creatures

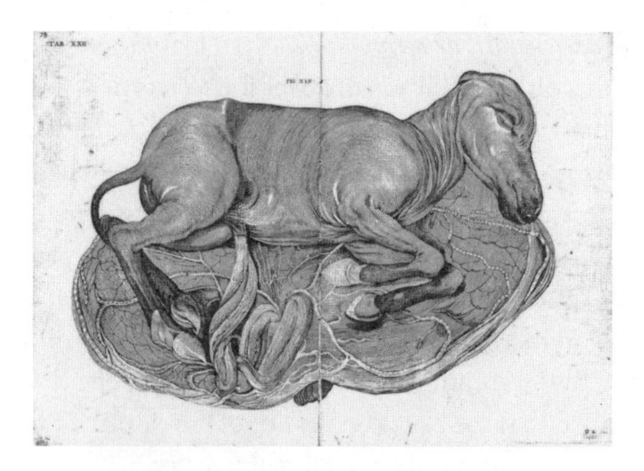

Fetal calf, from Hieronymus Fabricius ab Aquapendente, *De formato foetu*, 1600. Wellcome Collection

2.2 The Course of Gestation

In his work on generation, William Harvey describes a fetal deer:

About the end of December the foetus is a span long, and I have seen it moving lustily and kicking; opening and shutting its mouth; the heart, inclosed in the pericardium, when exposed, was found pulsating strongly and visibly; its ventricles, however, were still uniform, of equal amplitude of cavity and thickness of parietes; and each ending in a separate apex, they form together a double-pointed cone. . . . The internal organs, all of which lately had become perfect, were now larger and more conspicuous. The skull was partly cartilaginous, partly osseous. The hooves were yellowish, flexible, and soft, resembling those of the adult animal softened in hot water.

■ Quote from William Harvey, *Anatomical Exercises on the Generation of Animals* (1651), trans. Robert Willis (London: Sydenham Society, 1847), 494–96.

differed fundamentally from nonliving things. That difference could not be analyzed or seized by experimental art. The ultimate cause of life was unknowable, and the best science merely described its effects. To Harvey, the heart was simply an intermediate cause, not the secret of life, and the ultimate cause of generation could not be found by humans. The blood was for him the critical element as the agent of vital heat, the very essence of life. The blood was the first to live and the last to die, and Harvey called it "the immediate instrument of the soul," bearer of the vital principle endowed by God. The purpose of circulation was to revivify and reheat the blood, which cooled as it moved through the body.

Harvey concluded that life is autonomous and spontaneous and cannot be reduced to mechanics, to matter and motion. Nonetheless, his concept of the heart as a pump was such a compelling mechanical image that Harvey's interpretation, resisted by many of his Galenist contemporaries, was seized upon most of all by emerging mechanical philosophers, who believed that all phenomena were reducible to mechanics and saw no need for mysterious vital functions.

Descartes, the Villain of History

Despite Harvey's sacrifice of hundreds of animals, modern antivivisectionists and animal rights activists have ignored him. The villain's role in the story of the use of animals in science has instead been assigned to Harvey's younger contemporary the French natural philosopher René Descartes (1596–1650). In 1982, a member of the radical Animal Liberation Front slashed a portrait of Descartes at the Royal Society in London. The Australian philosopher Peter Singer, author of *Animal Liberation* (1975), calls Descartes's ideas about animals the "absolute nadir" of Western thought on that topic. In *The Case for Animal Rights* (1983), the American philosopher Tom Regan spends an entire chapter attacking Descartes's views. "It is tempting," he admits, "to dismiss Descartes's position . . . as the product of a madman."[3]

All this obloquy is directed at a man who performed very few experiments on animals, who liked his pet dog and, as far as we know, did not dissect him. Descartes's villainous reputation stems from his philosophical claim that animals were not conscious and therefore lacked the ability to suffer and feel pain in the way humans do. A modern medical ethicist repeats a widely believed view that "Descartes's denial that animals (despite all appearances to the contrary) were able to suffer, appears to have been widely used as a justification for experimenting on live animals, at a time when that practice was becoming more common."[4] But most of his contemporaries did not accept Descartes's views, which were not as straightforward as we have been led to believe.

Descartes was the chief theorist for the new mechanical philosophy. The central premise of this philosophy as it developed during the seventeenth century was that the universe was a machine that operated according to the laws of mechanics. In 1543, the same year Vesalius published *De fabrica*, the Polish astronomer and priest Nicolaus Copernicus (1473–1543) published his *On the Revolutions of the Heavenly Spheres* (*De revolutionibus orbium coelestium*), in which he proposed that the sun, and not the earth, was the center of the universe and that the earth was therefore one planet among many. The implications of this idea were only gradually realized. By the end of the sixteenth century, Galileo Galilei (1564–1642) and others recognized that if the earth was no longer at the center of the universe, then all of physics required rewriting.

While Galileo focused on the problems arising from Copernicus's astronomy, Descartes intended to systematize Galileo's mechanics and his mathematical approach into a new philosophy. This mechanical philosophy separated the psychic and spiritual world from the physical and material world and established the primacy of material (rather than other sorts of) causation. Matter itself was not active but inert. Theologians, such as Descartes's Jesuit teachers, feared that a clockwork universe governed by the laws of motion could operate without God's intervention. Descartes's philosophy of nature would, he thought, incorporate the new mechanical worldview while keeping intact God's central role in the governance of nature. Galileo's works were condemned by the Catholic Church, and he was placed under house arrest in the 1630s. Descartes hoped to avoid that fate.

While traveling as a soldier in the winter of 1619, Descartes spent a rare day alone in thought and developed the basis for his new philosophy. He later described this remarkable day's thoughts in his *Discourse on the Method* (*Discours de la méthode*, 1637). Descartes attempted to empty his mind of all preconceived ideas and to find what he knew to be true instinctively or intuitively—what appeared to him to be "clear and distinct" ideas. He discovered his thinking mind and therefore himself, who possessed the mind: "I think, therefore I am" (in Latin, *cogito ergo sum*). Descartes existed primarily, then, as a thinking being; his body was all but irrelevant. His second revelation was the idea of God who had made him, the idea of whom was so clear and distinct that it must be true. Descartes's third principle was the truth of geometry and mathematical relationships, which were self-evident *whether or not they referred to actual objects.* These self-evident principles were axiomatic and therefore incapable of analysis.

From this basis Descartes set out to reconstruct the world. He viewed the world as a collection of mechanisms that could be viewed and analyzed only in mathematical and mechanical terms. There was therefore no distinction between what we would call the physical and biological worlds, no special vital force. Everything could be analyzed in terms of the laws of mechanics. However, reality did not necessarily correspond to one's perception of it. To explain visible appearances in terms of matter in motion, he and other mechanical philosophers turned to the ancient doctrine of atomism, which stated that visible matter was composed of small particles individually invisible to the eye. The evidence of the senses, therefore, did not tell the whole story and could be misleading. Color and texture were merely the result of the disposition of the minute parts of matter.

Descartes proceeded, therefore, not by experimentation but by reasoning, and his discussion of the motion of the heart shows fundamental differences from Harvey. Descartes learned of Harvey's theory of circulation soon after its publication, and he probably witnessed a public demonstration of it in the Netherlands in the early 1630s. In *Discourse on the Method* he restated Harvey's discovery in mechanical terms. While Harvey had refused to assign a cause to the heart's motion, Descartes did not hesitate to assign a mechanical cause. Harvey's analogy of the heart to a pump was simply that—an analogy, not a description—but Descartes believed that the mechanisms he described were real.

Descartes did not describe the process of dissection or vivisection, but simply the result, for he would not demonstrate circulation through dissection, but by means of mechanical principles, by logic. The question of animal heat remained critical, but Descartes transferred its site from the blood to the heart, and the mechanical effects of the heart's heat on the blood gave rise to circulation. He did not measure that heat, and no numbers appear in his account. The heart's heat derived from fermentation, like that of wine, a mechanical process caused by the motion of small particles. It was not the result of a mystical vital force as Harvey contended. The heartbeat occurred when the heart's heat caused the particles of blood to expand and turn into vapor. The vaporized particles of blood then spewed into the arteries, causing the heart to contract and draw in more blood. The image is that of a boiling teakettle (constantly refilled with cold water) rather than a pump. Therefore, the dilating (diastolic) motion of the heart was the active phase of the heartbeat, the opposite of Harvey's view. Descartes claimed nonetheless that his description was consistent with observed phenomena. The configuration of the parts, he said, dictated this consequence just as the wheels and weights of a clock determined its motion.

Although the human body was a machine, humans also possessed rea-

Galatea fountain, from Salomon de Caus, *Les raisons des forces mouvantes*, 1615. Metropolitan Museum of Art

son. Descartes's first clear and distinct idea was that he was thinking, not that he had a body. Since humans could think, they necessarily had knowledge of God (Descartes's second clear and distinct idea), and therefore they possessed immortal souls. The essence of the world, to Descartes, was this dualism of mind and body, the complete separation of matter and spirit—a theological as well as a philosophical principle.

As Thomas Aquinas had been impressed with the new mechanical clocks of the thirteenth century, so Descartes marveled at the clockwork automata of his time, such as the mechanical fountains of Salomon de Caus (1575–1626). If humans could make such convincing devices, how much more clever was the hand of God, who had made the infinitely more complex animal machine. In *Discourse on the Method*, Descartes said that humans conceivably could make a mechanical animal that would be indistinguishable from the real thing. Nonetheless, a mechanical human could never be mistaken for a real human because it would lack a mind, which only God could bestow and which God bestowed only on humans. The mechanical human would manifest its inadequacy in two critical respects: it would lack speech, and it would lack the ability to reason.

Descartes believed that animals could neither speak nor reason and therefore were simply body, mere machines. The arrangement of their parts could lead animals to emit sounds in response to certain stimuli or to act in certain ways. God had constructed the body, said Descartes, to do quite a lot without reference to the mind. While he argued that animals were not self-conscious or capable of reasoned thought, they did, he believed, feel pain, heat, hunger, and emotions such as fear and joy. The mind's functions included memory, conscious perception, and most important, reason. As a Catholic who constantly sought the approbation of the Church, Descartes knew that Catholic doctrine did not grant animals an immortal soul. He had observed Galileo's fate in the early 1630s, and he so feared a similar punishment that he suppressed his major statement of a mechanical universe, *Le monde* (*The World*, 1633, published 1664). On the existence of animal intelligence and soul, Descartes sided with the Church against Michel de Montaigne (1533–1592), a skeptical philosopher who doubted the possibility of attaining true knowledge, and his disciple Pierre Charron (1541–1603). Montaigne argued that the complexity of spiders' webs and beehives implied that their makers possessed intelligence, even intelligence superior to humans'. In his "Apology to Raimond Sebond" (1580), Montaigne satirically claimed that animals surpassed humans in industry and in cruelty. Charron took this view more seriously and argued in his *Treatise on Wisdom* (*Traité de la sagesse*, 1601) that animals had souls.

Descartes dissected dead animals but seldom experimented on live animals. Unlike Harvey, Descartes did not regard experimentation as a method of discovery but saw it as a way to confirm what he had already deduced by mechanical principles. By deduction, Descartes could generate several plausible mechanical scenarios to explain a phenomenon. Experiments could help to determine which of these scenarios was correct. Many of Descartes's experiments served simply to extend observation, to display the true arrangement of the parts. While Harvey believed that the soul was contained within the body, Descartes (hewing far more closely to theological orthodoxy) believed that the soul was essentially different from the body and that learning about the body—especially the animal body—could reveal little about the soul.

"The Martyrs of Our Experiments": Animal Research in the Seventeenth Century

Descartes accepted Harvey's animal experimentation as a valid method of research, but conservative critics of Harvey argued that animal experimentation was useless.[5] They contended that experiments on animals could reveal nothing about humans because humans were unique. Some also revived the old empiricist argument cited by Celsus: that vivisection caused pathological changes in the research animal that invalidated the data. However, most researchers eagerly followed Harvey's methods, while rejecting his vitalist philosophy.

Almost all experimenters in the second half of the seventeenth century began their research with two fundamental assumptions: that organisms and machines were analogous and that all vital functions involved only the substances and processes of inanimate nature. There was no distinction between physics and physiology. The Cartesian notion that animals were machines (what came to be dubbed the "beast-machine") was far more important as a methodological principle, an approach to research, than as a moral principle or a description of what animals were really like. Researchers across Europe looked for evidence of mechanism in animal form and function, but they did not necessarily believe that the animals upon which they experimented felt no pain. At the same time, renewed study of natural history from the mid-sixteenth century onward, enhanced by animals previously unknown in Europe, led to a new appreciation for animals as animals.

As the royal physician, William Harvey followed King Charles I from London to Oxford in 1642 after civil war broke out. He remained there until 1646. In Oxford, Harvey found a congenial circle of men with whom he discussed his work and performed experiments. In 1660, members of this group formed a scientific club in London called the Royal Society, where they continued their experimenting. Their research revolved around problems Harvey had introduced with his discovery of circulation. These included the purpose of respiration, the cause of animal heat, how metabolism worked, why arterial blood was red and venous blood dark, and the cause of nervous transmission. However, they sought mechanical causes and not vital principles.

The Royal Society circle used many animals, alive and dead. Their techniques included the ligature of blood vessels, the inflation of lungs or other organs to show their structure, and the injection of various substances. Among the best-known experiments in this period were those employing a vacuum pump, or air pump, invented in the late 1650s by Robert Hooke (1635–1703) and Robert Boyle (1627–1691). Boyle and Hooke used the air pump to determine whether animals needed air to survive. Mice, cats, birds, snakes, and cheese mites were put in the vacuum or simply left inside the bell jar of the pump without withdrawing air from it. Boyle and Hooke concluded not only that animals needed air but that something in fresh air as opposed to old (respired or burned) air was essential to life. Air was a dynamic and necessary element in sustaining life.

Hooke and Richard Lower (1631–1691) undertook a series of surgical vivisections in the mid-1660s to determine the mechanism of respiration. In these experiments, the thorax and diaphragm of a dog were cut away, leaving its heart visible. The dog was kept breathing by pumping air into its windpipe through a bellows. As soon as the bellows stopped, the dog's heartbeat became irregular. While this experiment showed that air was essential to life, it could not disprove the claim that the motion of the lungs (as some scientists contended) was also essential. Hooke and Lower repeated this experiment a few years later, with a twist. They used two sets of bellows, which kept the lungs continuously full but motionless (they pricked the pleural membrane so that air escaped out the bottom). The heart continued to beat even when the lungs did not move, proving that the air, not the motion of the lungs, was the critical factor.

Some of the spectators at the Royal Society, while impressed with the results, expressed unease at this research program. John Evelyn, a fellow of the Royal Society, described the open-thorax experiment as "of more cruelty than pleased me."[6] He also disliked dogfighting, a popular spectator sport. Like many of his contemporaries, and few of his ancestors, he owned several dogs as pets. Hooke himself disliked the open-thorax experiment because of its cruelty, and it took three years for Lower to persuade him to repeat it. In his book *Micrographia* (1665), Hooke expressed his preference for the nonviolent observation of nature that the microscope afforded.

But Hooke continued to seek experimental demonstrations that could edify and entertain the Royal Society's membership of learned men and fashionable dilettantes. The technique of injection fulfilled the society's requirements of dramatic effects with serious intent. Researchers tried to trace

2.3 Pneumatick Engine

In 1659, Robert Boyle and Robert Hooke placed a lark in the receiver of their "pneumatick engine" and pumped out the air. The lark died, the first of many animals sacrificed to the air pump. Boyle's books and his papers in the *Philosophical Transactions* of the Royal Society, the first scientific journal, made the air pump widely known. This illustration of the second version of the air pump probably shows what Boyle called a "kitling" expiring in the vacuum. Hooke became curator of experiments at the Royal Society in the 1660s, and the air pump was a popular demonstration instrument at its meetings. It also proved to be an important step in the professionalization of science. The air pump was expensive and difficult to make, and it required special expertise to operate. Experiments that employed it could not be repeated by just anyone. This use of a machine may also have made it easier for researchers to think of the animals they used as mere machines as well.

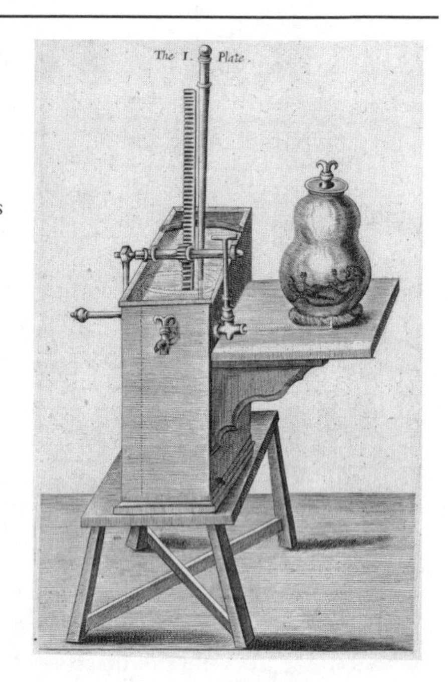

Robert Boyle, *A Continuation of New Experiments, Physico-mechanical, Touching the Spring and Weight of the Air, and Their Effects* (Oxford: printed by H. Hall for R. Davis, 1669). Wellcome Collection, CC BY 4.0

the course of blood vessels in animals with injections of ink. Others injected various drugs, poisons, and paralytic agents to show their mode of action. Attempts to feed intravenously with injections of broth and milk failed, but dogs got drunk when injected with wine and beer. The dramatic and the serious were not always equally balanced.

Entertaining, horrifying, and irresistible (to judge by the amount of publicity they received), the blood-transfusion experiments of the late 1660s dramatically displayed the strengths and weaknesses of the experimenters' program. The transfusion craze swept Europe. The composition of the blood was imperfectly understood, and the concept of blood types unknown. Many physicians believed that transfusion had great therapeutic potential. Even if experimenters did not agree with Harvey that the blood contained

2.4 New Dogs from Old?

In his 1676 play *The Virtuoso*, Thomas Shadwell satirized the Royal Society's best-known experiments. The top illustration shows a transfusion between a lamb and a man. In this passage from Shadwell's play, the virtuoso, Sir Nicholas Gimcrack, explains his experiments to the "gentlemen of wit and taste," Longvil and Bruce and his friend Sir Formal Trifle:

LONGVIL: Sir, I beseech you, what new curiosities have you found out in physic? . . .

SIR NICHOLAS: Why I made, sir, both the animals to be emittent and recipient at the same time. After I had made ligatures as hard as I could (for fear of strangling the animals) to render the jugular veins turgid, I open'd the carotid arteries and jugular veins of both at one time, and so caus'd them to change blood one with another.

SIR FORMAL: Indeed that which ensu'd upon the operation was miraculous, for the mangy spaniel became sound and the bulldog mangy.

SIR NICHOLAS: Not only so, but the spaniel became a bulldog and the bulldog a spaniel.

BRUCE: 'Tis an experiment you'll deserve a statue for.

Georg Abraham Mercklin, *De ortu et occasu transfusionis sanguinis*, 1679. Wellcome Collection

■ Quote from Thomas Shadwell, *The Virtuoso* (1676), ed. Marjorie Hope Nicholson and David Stuart Rodes (Lincoln: University of Nebraska Press, 1966), 47–48.

the soul, ideas lingered of its role as the repository of temperament. A German doctor suggested that transfusions between a husband and wife could lead to conjugal harmony. The central ritual of Christianity transformed wine into the blood of the Savior. Furthermore, despite Harvey's convincing proofs that the body could not generate an infinite supply of new blood, bleeding continued to be a common therapy.

Animal-to-animal transfusions ran the gamut, as blood was exchanged between "Old and Young, Sick and Healthy, Hot and Cold, Fierce and Fearful, Tame and Wild Animals."[7] At the Royal Society, Richard Lower and his colleagues began with dog-to-dog transfusions but soon tried cross-species transfusions with sheep and calves as well as dogs. Members of the new Royal Academy of Sciences in Paris soon followed. Inevitably, human transfusions followed. The experimenters argued that the blood of young calves and lambs was purer than human blood, less tainted by human passions and vices. Did not the blood of the lamb symbolize the pure healing blood of Christ? Healing the passions of the mind was an especially important goal of these early transfusers. Yet their overly passionate (if not quite insane) subjects were also in no position to complain about their treatment. In these early attempts at human experimentation, no notion of "informed consent" arose, and neither the patients nor the transfusers seriously questioned the morality of the experiment.

A French royal physician, Jean-Baptiste Denis, transfused the blood of a sheep into a young man suffering from a fever, whose own blood supply had been depleted by therapeutic bleeding. Denis reported in the *Philosophical Transactions* that the young man appeared recovered. Lower and a colleague soon tried a similar experiment at the Royal Society, transfusing sheep's blood into a young clergyman whose brain had been described as "sometimes a little too warm." The transfusion appeared to have the desired effect of calming the young man's behavior.[8] A second transfusion on the same individual a few weeks later brought out a crowd of spectators. Shortly thereafter Denis transfused a similarly manic patient, again reporting success. However, the patient died following an attempt at a third transfusion a few months later. Both the French and English patients complained of various symptoms after the second transfusion. They were experiencing a *hemolytic reaction* to the foreign blood, characterized by a massive die-off of their own red blood cells. That they did not experience reactions sooner can probably be attributed to the inefficiency of the transfusion process. It is unlikely that they received more than a few ounces of animal blood, but that would have been enough to stimulate antibodies to the foreign blood, and subsequent transfusions would have led to more severe consequences.

The death of Denis's patient early in 1668 quickly ended any further attempts at animal-to-human blood transfusion. Experimental blood transfusion only resumed in the 1820s, and human-to-human blood transfusion did not become an accepted medical practice until after the discovery of blood types in the early twentieth century.

Although the transfusion experiments attracted much public attention, they were only a small part of the research enterprise in this period, which was a fiercely contested territory of competing theories. Most researchers agreed that Harvey's theory of circulation was correct and that his anatomical methods were useful, but the interpretation of his discovery, as we have already seen in the case of Descartes, could vary widely. Boyle and his circle were mechanists in the sense that they believed the animal body could be analyzed in mechanical terms. But they did not follow Descartes in attempting to discover or hypothesize the micromechanics that made the body function. The Paris Academy's Claude Perrault (1613–1688) also disputed the beast-machine doctrine, asserting that animals had a level of consciousness and an immaterial soul. While national rivalries, especially between the French and the English, played a role in the transfusion experiments, new scientific journals also ensured that competing researchers shared their ideas with one another and with the interested public.

The Italian anatomist Marcello Malpighi (1628–1694) and his colleagues explored another variation on the animal machine in their animal experiments. Rather than physiological processes such as respiration, Malpighi investigated the microstructure of the lungs and other organs. Over a period of years, he and his colleague Carlo Fracassati (d. 1682) dissected and vivisected guinea pigs, cats, sheep, frogs, and birds. Malpighi injected lungs and pulmonary vessels with colored water to determine their structure. The lungs were revealed as a series of small cellular vesicles separated by membranes, rather than solid or even spongy flesh. More difficult was the tracing of the pulmonary circulation. No one had yet seen the anastomoses, or connections, between the smallest veins and capillaries, even though Harvey's theory required them. Malpighi's injection methods encountered technical problems: in dead animals the fine vessels were clogged with clotted blood, while in live animals the blood flowing through the vessels could not easily be replaced with colored water without killing the animal.

Malpighi's correspondence with his friend Giovanni Alfonso Borelli

(1608–1679) over a period of two years reveals the hit-and-miss nature of experimentation: one method failed, he tried another, Fracassati reported different results, Borelli made suggestions, Malpighi tried again, a different animal, a different technique. The anastomoses must exist. But how to prove it? Fracassati and then Malpighi tried frogs. Borelli complained that he did not have access to frogs and could not duplicate Malpighi's observations. Malpighi nonetheless continued, sacrificing, he said, almost the entire race of frogs.[9]

Frogs were excellent subjects for this research: their lungs are relatively simple in structure and nearly transparent, allowing a clear view of the blood vessels within. Malpighi observed the living frog and then tied off the lungs top and bottom, retaining the blood within. He dried the tiny lungs, inflated them, and then examined them with a microscope. There he found the connections he sought. His remarkable treatise on the lungs, *Epistolae de pulmonibus* (1661), gave new prominence to the recently invented microscope as a research tool. The microscope revealed the minute subtlety of nature, which atomism and mechanical theories only surmised. Malpighi hypothesized that the lungs functioned to break the blood down into ever more minute particles that would then circulate through the body to become bone, organ, or fluid. Nutrition was not mysterious, but a purely mechanical process. He rejected aspects of Cartesianism as too speculative, yet he regarded the animal as a machine.

By the end of the seventeenth century, the animal body (and by analogy the human body) was better known than it had been at any time in the past, and scientists pointed to their own successes to justify their experimental practices against critics. The encyclopedic *Bibliotheca anatomica* (Anatomical library, 1685) proclaimed the triumph of the mechanical philosophy. Robert Boyle defended the practice of natural philosophy, including animal experimentation, in his *Some Considerations Touching the Usefulness of Experimental Natural Philosophy* (1663). A devout Christian, Boyle refuted claims that science threatened religion, referring to the "two books" idea that contemplation of nature led to a finer appreciation of God through his works. Boyle also revived the old concept of stewardship. God had entrusted the earth to humans to improve it by their activities, not merely to contemplate it. Boyle believed that animal experimentation provided an excellent example of divine sanction. Not only did it afford knowledge of God's creation and its purpose to the experimenter but it also provided

2.5 Beast-Machine

The Swiss physician Albrecht von Haller (1708–1778) experimented on dogs, attempting to demonstrate the sensations created by various forms of stimuli, including painful ones. Clearly, Haller believed that these dogs felt pain. Only the most fervent Cartesians embraced the "beast-machine" concept so far as to act out its consequences. In the seventeenth century, this included the religious order of Jansenists, centered at the Parisian monastery of Port-Royal. A secretary to the Jansenist fathers described their cruelty to animals in a much-quoted passage:

> They administered beatings to dogs with perfect indifference, and made fun of those who pitied the creatures as if they had felt pain. They said the animals were clocks; that the cries they emitted when struck, were only the noise of a little spring which had been touched, but that the whole body was without feeling. They nailed poor animals up on boards by their four paws to vivisect them and see the circulation of the blood which was a great subject of conversation.

Title page from Albrecht von Haller, *Memoires sur la nature sensible et irritable, des parties du corps animal*, vol. 1, 1756–60. Bibliothèque interuniversitaire de santé, Paris, Licence ouverte

Another Cartesian clergyman and philosopher, Nicolas Malebranche (1638–1715), reportedly kicked a dog at his feet and responded coldly to a protesting observer, "So what? Don't you know that it has no feeling at all?" Malebranche's denial of animal suffering was theological. Human suffering, including that of children, could be attributed to the original sin of Adam, which condemned all of humanity. Since animals were not descended from Adam, they could not suffer, for God would not have created pointless suffering.

■ Both quoted in Leonora Cohen Rosenfield, *From Beast-Machine to Man-Machine* (1940; reprint, New York: Octagon, 1968), 54, 70.

useful knowledge. To Boyle, it was obvious that God had provided animals to the anatomist so that he could conduct experiments he could not pursue on humans.

Boyle and his colleagues were aware that the animals could suffer. Although most researchers adhered to some version of the mechanical philosophy, and many were convinced Cartesians, few of them took literally Descartes's argument that if animals were machines, they did not feel pain as humans did. Boyle noted an animal's distress during an experiment and reported that a viper had been "furiously tortured" under the influence of the vacuum.[10] Lower bled animals to the point of death, determining this point from the animals' struggles. While a Cartesian might argue that these struggles were purely automatic, Lower did not argue this, instead employing language that denoted pain and suffering. Carlo Fracassati injected a dog with vitriol (hydrochloric acid) and noted, "The Animal complain'd a great while . . . and observing the beating of his breast, one might easily judge, the Dog suffered much."[11] All of them believed that such suffering was preferable to human suffering for the sake of advancing knowledge.

Few seventeenth-century scientists used the supposed insensitivity of animals as an excuse for experimenting on them. A clergyman-scientist commented to a friend, "I wish [the Cartesians] could convince me as thoroughly as they are themselves convinced of the fact that animals have no souls!!"[12] Rather, the concept of the "beast-machine" described a highly successful approach to research. Some natural philosophers and observers in this period nonetheless began to experience pricks of conscience at animal experimentation.

Beginnings of Moral Concern

Galen wasted no thought on moral considerations when he cut open living animals, and in this he was a man of his time. By the time Vesalius duplicated some of Galen's experiments in the 1530s, social standards had changed. In the Renaissance the humanist ideal of a cultured individual was superseding the warrior ideal of the medieval knight. The path to success in the sixteenth century was as much at the royal court as on the battlefield, and skills in polite conversation were valued. The ideal humanist was also a Christian who would display kindness and compassion.

Vesalius seems little different from Galen in his feeling for animals. But the spectacle of public anatomy, which included demonstrations on live animals, was intended (unlike in ancient Rome) to have a moral impact on its audience. Viewing the end of all flesh reminded the audience of the importance of their souls, and suffering animals induced compassion. In the finale of his public anatomy, Vesalius vivisected a pregnant animal, either a sow or a dog, expertly manipulating both the animal and his audience.

Vesalius's successor as professor of anatomy at Padua, Realdo Colombo, cut open a pregnant dog, removed the puppies, and then hurt them in front of the mother. Ignoring her own pain, she tried to comfort the pups. Colombo reported that the bishops and other clergymen in attendance were especially impressed by this display of motherly love. The juxtaposition of life and death, pain and pleasure, had a powerful emotional impact, and the fact that the sufferer was animal rather than human did not diminish this impact. In public anatomy, animals acted as moral and physical proxies for humans. Thomas Aquinas had declared that being cruel to animals could lead to being cruel to humans. But Thomas did not condemn the suffering of the animal as morally objectionable in itself.

In the seventeenth century, as the use of animals in experiments increased, popular attitudes toward animals also began gradually to change. Both public anatomy and the use of animals in popular entertainments such as bearbaiting and dogfights continued throughout the eighteenth century. But public anatomy had disappeared by 1800, when animal sports (with the notable exception of hunting) were increasingly viewed as lower-class activities, too disturbing for educated tastes. Meanwhile, petkeeping increased: the English monarch Charles II's mistresses had their portraits painted with their lapdogs, and a popular English pamphlet of 1644 bemoaned the death of a pet dog in battle. The French noblewoman Madame de Lesdiguières built an elaborate tomb for her cat Ménine in 1684.

Public knowledge of scientific practices meant that they were not immune from public criticism. In France, the notion of the "beast-machine" came in for strong criticism. In his *Discourse on Animal Knowledge* (1672), the Jesuit priest Ignace-Gaston Pardies relied on Aristotle to support his argument that animals could feel, imagine, and remember, although they could not contemplate or have spiritual knowledge. Moreover, Pardies continued, God would not have given animals sense organs just for show. In 1690, another Jesuit, Gabriel Daniel, extended Pardies's arguments in his *Voyage to the World of Descartes*. Daniel cited animal experimentation as an instance of extreme cruelty, arguing, as Thomas Aquinas had, that cruelty to animals would lead to cruelty to humans. Daniel did not, however, argue that animals had rights, a notion that would have seemed absurd to him.

John Ray (1627–1705), an English clergyman and naturalist, attacked the "beast-machine" notion on theological and scientific grounds. In a 1693 essay, Ray argued that God had not created the world for humans alone. Animals existed of themselves, not merely for human use, and expressed God's creative power. He argued that animals were conscious but not rational. This argument did not mean that treatment of animals should be given moral consideration, but Ray nonetheless believed that animal suffering was immoral, declaring, "The torture of animals is no part of philosophy."[13]

Harvey's discovery of circulation brought animal experimentation to the forefront as a scientific method. The mechanical philosophy of nature, most forcibly enunciated by Descartes, gave an added dimension to this method as researchers sought mechanical explanations for the operations of the human and the animal body. While some followed Descartes in seeking actual micromechanisms, others viewed mechanism as a convenient metaphor. Few believed Descartes's contention that animals' lack of consciousness meant that they could not in some way feel pain. The increased number and invasiveness of experiments led some to consider animal suffering as a moral issue for the first time since antiquity.

3 Disrupting God's Plan

Shortly before Christmas 1694, Queen Mary II of England, aged 32, died of smallpox. Her death set in motion a chain of events with profound dynastic consequences for the English crown. Mary and her Dutch husband, William of Orange, had assumed the English throne six years earlier when her father, King James II, was deposed. Mary and William had no children, and her death in the middle of her childbearing years meant that the heir to the throne was now her sister Anne, whose only son, the Duke of Gloucester, then aged 5, was not at all healthy. When the young duke died of smallpox in 1700, Parliament took the question of the succession into its own hands. The Act of Settlement in 1701 placed the line of succession in a set of German cousins rather than allowing James or his children from his second marriage to return to England. The first of the kings from Hanover, George I, took the throne in 1714. He spoke no English.

Thus could disease determine the fate of dynasties and nations. Although in this age before antibiotics many infectious diseases were endemic, smallpox was an especially omnipresent affliction. The virus that caused it could travel through the air, making it highly contagious, and one might contract the disease simply by being in the same room with a smallpox sufferer. Its course included a high fever, upper-respiratory symptoms, and finally the characteristic "pocks," or rash. Each stage of the disease carried its own dangers, but the pocks were the most dangerous and most feared. Discrete pocks were less dangerous than confluent ones that ran together into one suppurating sore, but in either case the pocks could attack the eyes as well as the inside of the mouth and throat. What we now recognize as different strains of the smallpox virus could cause more or less severe symptoms. How smallpox, particularly the more virulent variola major, reached Europe is uncertain, but by the sixteenth century it had become established as one of the more serious of the many hurdles children faced in their struggle to reach adulthood. Adulthood did not in any case offer safety. The pockmarked faces and bodies of many adults and the blindness of others gave evidence of that struggle. Some did not survive the disease. In an average year about 20 percent of all those infected died; in years of epidemic a larger percentage perished. During the epidemic of 1721 the parish of St. Giles-in-the-Field in London recorded 68 deaths during the week of 14 March. Smallpox had claimed 19 of those.

Prevailing medical theories offered the smallpox victim conflicting advice. While the disease was obviously contagious, no one knew exactly how sufferers conveyed it to others. Especially in its early stages, smallpox was difficult to distinguish from other rash-producing diseases such as measles. Surviving an attack was believed to confer immunity, but no one knew how or why this occurred or how long it lasted. The French king Louis XV was diagnosed with a mild case of smallpox in 1728 and was afterwards thought to be immune. When he became ill in 1774, smallpox was not suspected until far too late, and Louis died miserably. But what, in any case, could his doctors have done? Some physicians advocated keeping the patient warm, while others urged a cooling therapy to counteract the effects of fever. Amulets were used to ward off the disease, and noxious potions of tar water or sheep's dung were used to cure it. Patients were bled and purged, although doctors disagreed about the correct time for these procedures. Some still advocated the medieval "red therapy" of surrounding the patient with the color red to "draw out" the pocks and thus draw the disease out of the body. England's Queen Elizabeth I had supposedly been cured of an attack in 1562 by a combination of two therapies: she was wrapped in red flannel, and she was placed before the fire to keep warm. Her royal descendant Mary II was less fortunate.

Europeans had brought smallpox and other diseases to the New World, where it decimated indigenous populations. Many in Europe believed it was increasing in severity in the seventeenth and eighteenth centuries, and epidemics such as in London in 1681 seemed to confirm this. Among childhood diseases, it had a high rate of mortality (as much as 40 percent in some years), and survivors endured a lifetime of disfigurement. It was among the most loathed and feared, and most common, of diseases. These factors led many Europeans to participate in the largest human experiment ever conducted: the smallpox inoculations of the eighteenth century. The inoculation controversy raised several issues that would reappear in later discussions of human experimentation, particularly in twentieth-century clinical trials. These included the concept of acceptable risk, the amount of information the patient should have, and the physician's responsibility to the community.

Most agreed that those who had once had smallpox did not usually get it again. Some parents would allow their uninfected children to catch the disease from their siblings, and in rural areas children were often urged to get in bed with relatives suffering from smallpox. In other areas, "buying the pox" was also common. Parents would send their children to the home of a victim of a mild attack to purchase a few of the scabs fallen from the pustules. African slaves reported similar practices in their homelands of deliberately inducing the disease to confer immunity. But no one understood how the disease was contracted or how immunity was conferred.

Reports from outside Europe of inoculation, of deliberately inducing smallpox to confer immunity, were received by the Royal Society of London, the most famous scientific society in Europe. In 1700, a physician described to the society the Chinese practice of inhaling dried smallpox matter to induce the disease. In 1714 and 1716, accounts appeared in the society's *Philosophical Transactions* of the Turkish method of inoculation, inserting matter from smallpox sores into small cuts in the skin, usually in the arm. This resembled African practices and was widespread in the Middle East.

European physicians were reluctant to adopt inoculation even though it was apparently successful. Its folk origins and associations with "wise women" made it suspect. In addition, physicians questioned whether a treatment that worked in a hot country would work in other climates and on other peoples. Despite long years of use elsewhere, the practice remained untested in Europe. Physicians were notoriously conservative in their therapies, as was evident in the continued use of bloodletting, even after Harvey's theory of circulation would, it seems, have made the practice illogical. And what physician who valued his reputation would deliberately induce a disease? The general sentiment was that inoculation was experimental in every sense of the term, and no physician wished to be the first to perform the experiment on his patients.

Two individuals who did much to promote smallpox inoculation in the early 1720s were not physicians. Cotton Mather (1663–1728), in New England, was a well-known Puritan clergyman and natural philosopher. Lady Mary Wortley Montagu (1689–1762) had witnessed inoculation in Constantinople as the wife of the British ambassador there from 1716 to 1718, and the embassy's surgeon, a Scot named Charles Maitland, had inoculated her young son in 1718. The forthright Lady Mary disliked and mistrusted physicians. When an epidemic of smallpox attacked London in 1721, she called on Maitland to inoculate her 3-year-old daughter, also named Mary. Maitland nervously complied, even though Lady Mary had refused to allow any physicians to be present as witnesses. This was the first recorded inoculation in England. Several physicians examined young Mary after the inoculation. She came down with a mild case of smallpox and survived. Everyone knew that this was literally a shot in the dark. Little Mary Montagu came through splendidly, but the 2-year-old son of the Earl of Sunderland, inoculated a year later, came down with a severe case of smallpox and died. Yet smallpox engendered such fear and horror that parents made their own cost-benefit analysis and took the risk. This was indeed an age of risk-taking; the wars of the seventeenth century had ended (only temporarily, as it turned out), but the new stock exchanges of London and Paris allowed a different kind of risk, and the South Sea Bubble of 1720 took investors on a dizzying ride from the heights of wealth to the depths of poverty after it crashed.

Almost despite themselves, the physicians who examined young Mary Montagu were impressed. One prevailed upon Maitland to inoculate his own 6-year-old son. Sir Hans Sloane, physician to the king, wanted experimental proof, even though the very act of inoculation could be seen as an experiment, since no one could be certain of the outcome. Such an experiment could only be conducted on humans, because only humans could get smallpox. Sloane turned to the most powerless people in society: prisoners. He asked the king for permission to inoculate several condemned prisoners in London's Newgate Prison. The prisoners would volunteer their services in exchange for their release—if they did not die from the experiment. The king gave his permission. The British royal family, perhaps mindful of the past dynastic consequences of smallpox, was among the first to hear about Mary Montagu and was quite interested in inoculation.

On 9 August 1721, Sloane recruited Charles Maitland to inoculate six prisoners, three male and three female, aged 19 to 36, before a crowd of learned witnesses. Maitland inoculated them, but a few days later, dissatisfied with the appearance of the incisions, he repeated the operation with fresh infectious material. Closely watched, five of the six came down with

3.1 The Turkish Practice of Inoculation

Lady Mary Wortley Montagu described the Turkish practice of inoculation in a letter to a friend:

The Small Pox so fatal and so general amongst us is here entirely harmless by the invention of engrafting (which is the term they give it). There is a set of old Women who make it their business to perform the Operation. Every Autumn in the month of September, when the great Heat is abated, people send to one another to know if any of their family has a mind to have the small pox. They make partys for this purpose, and when they are met (commonly 15 or 16 together) the old Woman comes with a nutshell full of the matter of the best sort of small-pox and asks what veins you please to have open'd. She immediately rips open that you offer to her with a large needle (which gives you no more pain than a common scratch) and puts into the vein as much venom as can lye upon the head of her needle, and after binds up the little wound with a hollow bit of shell, and in this manner opens up 4 or 5 veins. . . . The children or young patients play together all the rest of the day and are in perfect health till the 8th. Then the fever begins to seize 'em and they keep in their beds 2 days, very seldom 3. They have rarely above 20 or 30 in their faces, which never mark, and in 8 days time they are as well as before their illness. . . . Every year thousands undergo this Operation, and the French Ambassador says pleasantly that they take the Small pox here by way of diversion as they take the Waters in other Countrys.

Lady Mary Wortley Montagu. Lithograph by A. Devéria after C. F. Zincke. Wellcome Collection

■ Lady Mary Wortley Montagu to Sarah Chiswell, 1 April 1717, in Lady Mary Wortley Montagu, *Selected Letters*, ed. Isobel Grundy (London: Penguin, 1997), 158–59.

smallpox; one, it was discovered, had already had the disease. All of them recovered and were released.

The London newspapers seized upon the Newgate experiments with alacrity. Some reprinted the accounts from the *Philosophical Transactions*. Others expressed skepticism; *Applebee's Journal* noted that inoculation was "an Invention that had its Rise among the Populace, who were neither Men of Fortune, Character, nor Learning." Even after the apparently successful conclusion of the experiment, *Applebee's* remained doubtful: "Any Person that expects to be hang'd may make Use of it."[1]

Several newspapers and pamphlets pointed out that Sloane had not proven that inoculation conferred immunity. In response, he sent one of the released prisoners, 19-year-old Elizabeth Harrison, to nurse a smallpox victim under Maitland's surveillance. Even though she had fulfilled her side of the experiment, her class and perhaps also her gender made her vulnerable to further exploitation. The consent of Elizabeth Harrison, who was illiterate and unemployed, from inside and outside of prison was not informed and only marginally voluntary. Maitland reported to Sloane that Harrison "lay in the same bed" with her patient "every other night"; when a young boy also came down with the disease, Maitland "obliged [Elizabeth] to ly every night with the boy; and to attend him constantly from the beginning of the Distemper to the very end; And thus, she continu'd for six weeks together without Intermission; or suffering the least Head-ach, or other Disorder; tho, indeed, she once had some heats and Little pimples; as Nurses commonly have under such Confinements."[2] There is no record of Harrison's opinion of all this.

Harrison's continued good health persuaded a few more doctors that inoculation did indeed confer immunity, and they in turn began to inoculate their patients. In November 1721 the Princess of Wales (who had herself survived smallpox) announced that she would pay for the inoculation of all the orphans in St. James's Parish who had not yet had smallpox. This magnanimous gesture was actually another experiment to test the effects of inoculation on children. Although this grand plan fell through, Maitland did inoculate at royal expense six more volunteers, who were then available to be examined by the public, and in the spring of 1722 five orphans from St. James's Parish were inoculated and similarly displayed. The Princess of Wales was convinced, and in April 1722 Maitland inoculated her two daughters, the princesses Amelia and Caroline.

Royal validation led to a flurry of inoculation among the British aristocracy. Yet many questions remained unanswered: How did inoculation work? Why did some of those inoculated get mild cases of smallpox and others severe ones? How long did this induced immunity last? Couldn't inoculated victims spread the disease as readily as natural cases (they could, and did)? Could physicians justify causing disease, for whatever reason? The activities of Maitland and Sloane took place within a narrow circle of people and on a small scale, and at first they raised little opposition. Sloane quickly assumed leadership of inoculation with the Newgate trials. He was first physician to the king, president of the Royal College of Physicians of London, and the first physician to be named a baronet (a hereditary knighthood). As he was the most powerful physician in London, his advocacy of inoculation stilled many critics, at least for a time. At the same time in New England, however, inoculation was attempted on a much larger scale and aroused furious debate that soon reached the home country.

In Boston, the clergyman and natural philosopher Cotton Mather, fellow of the Royal Society, had learned of inoculation from his African slave Onesimus and read about it in the *Philosophical Transactions.* When a smallpox epidemic struck Boston in the spring of 1721, Mather encouraged local physicians to try inoculation. Unlike Lady Mary, who had Maitland at her side as well as her own son for corroboration, Mather had only his own knowledge and authority as a prominent clergyman. Boston's physicians, led by Dr. William Douglass, were not convinced. Only Dr. Zabdiel Boylston (1679–1766) responded, inoculating his sons, several of his slaves, and a few patients. A large public outcry soon followed against Mather, Boylston, and inoculation. Some argued that it only served to spread the disease by giving it to people who might not otherwise have acquired it. William Douglass condemned Mather and Boylston for their "rash and thoughtless Procedure in a *Medical Experiment* of Consequence," but he held Mather especially responsible as an amateur usurping the physician's role.[3] Among religious Bostonians, another argument was even more compelling: inoculation implied a lack of trust in God's overriding plan, amounting to an attempt to supersede God's authority. It was, after all, a heathen invention. In disregarding God's benevolent providence, its practitioners threatened to stir up God's wrath, which had already been manifested in the epidemic itself. Many argued that Bostonians should be praying and begging for forgiveness rather than tempting God further.

A pamphlet war raged for several months, but Boylston continued to inoculate patients, including Mather's son. While Boylston and Mather ultimately won the war of public opinion—only 12 of the 400 Boylston had inoculated (1 in 33) died, compared with 500 of the 3,600 natural cases (1 in 7)—they played a dangerous game, ethically and medically. Since no one understood how inoculation worked, the severity of the induced cases was largely a matter of luck, and critics were not entirely wrong in accusing Mather and Boylston of spreading the disease. Inoculation created new cases of smallpox, and those inoculated tended to act as if they were not in fact infectious, dismissing precautions against further spreading the disease.

In England too, pamphlets began to fly. Reports of inoculation from the *Philosophical Transactions* were repeated, not always accurately, in the popular press. Clergymen declared that only God had the power to give disease and to take it away. Opposition also rose among medical men, uneasy at the notion of causing disease and uncertain of the nature of infection. How did Sloane and his colleagues know that smallpox was indeed a single disease? Disease theory, based on humors, continued to focus on individual imbalance rather than infectious agents, and many physicians remained unconvinced that the disease induced by inoculation was truly smallpox. Some also worried that the smallpox matter could be contaminated with other diseases—which was indeed often the case. Moreover, who would wish to trust the judgment of "a few *ignorant women*"?[4]

Douglass in Boston argued that inoculation, like blood transfusion, was a parlor game of natural philosophers and should not be attempted on the general population. Many of the natural philosophers in the Royal Society supported inoculation, and they answered its critics with the authority of mathematics. While Mather, like the ancient empirics, argued that experience would be the most certain proof against dogmatic critics, physician-scientists such as James Jurin (1684–1750) employed numerical arguments. The absence of a convincing medical theory for the success of inoculation gave numerical arguments even more weight. Intellectuals in this age of Enlightenment firmly believed that to mathematize a problem was to explain it, and because numbers were believed to be innately true, a mathematical explanation was also true. Their model was the great English scientist Sir Isaac Newton (1642–1727), president of the Royal Society from 1703 to 1727, whose *Mathematical Principles of Natural Philosophy* (1687, known as the *Principia* from its Latin title) offered a mathematical explanation of the operation of the universe. His sophisticated mathematics explained motion both in the heavens and on earth, fulfilling Galileo's aim of unifying the natural world by means of mechanics. Newton's scientific method, including his use of mathematics, his use of experiments (particularly in the *Opticks* [1704]), and his emphasis on effects rather than causes, was best illustrated in his discussion of universal gravitation. He demonstrated this concept experimentally, proved its existence mathematically, and refused to speculate on its cause. To demonstrate its existence was sufficient; unlike Aristotle, Newton did not seek ultimate causes and indeed found them irrelevant to the pursuit of natural philosophy. His method and his discoveries influenced all areas of scientific endeavor in the eighteenth century.

James Jurin was the secretary of the Royal Society in the 1720s. He used his wide network of correspondents to survey the effectiveness of inoculation by comparing the number of deaths caused by inoculation with the number of those caused by naturally induced smallpox. Gathering case histories from a wide range of practitioners over a period of several years, Jurin collated this evidence and provided convincing proof that the mortality rate from natural smallpox far exceeded that from the inoculated version. Jurin analyzed the London bills of mortality (the official reports of deaths in the metropolis) for a 20-year period to determine that the risk of dying from natural smallpox was about 1 in every 8 cases; in times of epidemic, this could be 1 in 5 or 6. By contrast, 1 in 60 had died among inoculated cases in New England, and Jurin arrived at a similar figure for the period of 1721 to 1727 in England.

Jurin's numerical arguments, based on the emerging science of probability, persuaded those who sought answers in scientific authority. His arguments supported Newton's model of an orderly and balanced universe in which natural phenomena followed laws and were predictable. God had designed this universe—not a wrathful God who punished sin with disease but a God who gave humans the tools to conquer disease. The age of miracles was over. Newton's ideas gradually spread across Europe, and numerical arguments aided in the adoption of inoculation in other countries, particularly France, where physicians at first strongly resisted it. Daniel Bernoulli (1700–1782), a mathematician and member of the Paris Academy of Sciences, analyzed smallpox mortality in 1760, arguing that out of 13,000 children, inoculation would save the lives of 1,000 who would otherwise die of smallpox. By the end of the eighteenth century, inoculation was widespread in Europe, and itinerant inoculators, such as the Sutton family in England, inoculated thousands of people. Inoculation was risky, but taking one's chances with natural smallpox was far riskier. The mathematics of probability that Jurin and Bernoulli used was first developed in the seventeenth century to analyze games of chance, to predict who would win a bet. Inoculation was surely the biggest gamble of the eighteenth century, and it remained essentially an experiment, since no one knew how it worked or why.

Mathematical analysis entered every science, including social science, in the eighteenth century, and inoculation was one of the first scientific uses of statistics, now so central to scientific practice. Jurin boiled his case histories down to a stark numerical narrative. The individual patient became simply a number, all personal details effaced. This was an important step in the recognition of individual, identifiable diseases that acted the same in all victims, as opposed to the older view of diseases that differed according to individual temperaments. However, it also was a step in the depersonalization of the patient into a statistic, an experimental subject rather than an individual. This happened first with humans rather than with animals: until well into the nineteenth century, the number of animals used in any one experiment was far too small to allow for statistical inference.

Vile Bodies

Hans Sloane's decision to test inoculation on condemned prisoners had ample precedent. In antiquity, Herophilus and Erasistratus had experimented on prisoners, as had the "poison king," Mithridates. The medieval emperor Frederick II also tested poisons on prisoners and may have conducted other experiments on them as well. In the sixteenth century, certain physicians and pharmacists conducted trials of poison antidotes such as the "bezoar stone" in both animals and human prisoners. In seventeenth-century France, the physician Adrien Helvétius (1662–1727) obtained permission from King Louis XIV to test his secret remedy for dysentery on patients from the Hôtel Dieu, a large Paris hospital. Hospitals in this period were charitable institutions for the poor.

The historian Grégoire Chamayou employs the term *vile bodies* to refer to people who had become experimental subjects, including in this group prisoners, orphans, prostitutes, people with disabilities, the mentally ill, hospital patients, slaves, and the colonized.[5] To these we might add certain racial and economic groups and soldiers. How one became a vile body shifted with time and circumstance. The Princess of Wales felt that orphans could serve as inoculation test cases in 1721. Yet utility provided an overarching reason for experimenting: the princess claimed that inoculating the children would be beneficial for them. In the 1760s and 1770s, the physician John

Quier inoculated hundreds of slaves on plantations in Jamaica. By this time, inoculation of slaves, which had been suggested as early as 1722, had become commonplace. Quier tried various techniques, including inoculating people who had already had smallpox, as well as infants and pregnant women, and he sent his detailed notes to London for publication. Motivated by utility, Quier tested the boundary between therapy and exploitation.

Soldiers and sailors, captive groups in service to the state, provided another arena for experimentation. Scurvy, a disease caused by a deficiency of vitamin C, which caused weakness, fever, fatigue, and ultimately death, was widespread among sailors in the eighteenth century, who undertook long voyages on a diet mainly of hard biscuits and salt pork. James Lind (1716–1794), a ship's surgeon in the British navy, believed that this diet caused the disease. In 1747 he enlisted 12 sailors with scurvy to test his theory; he divided them into six pairs, who ingested differing remedies, such as sea water, cider, and vinegar, over 14 days. The best result by far came from oranges and lemons. Lind described this trial in his 1753 *Treatise on the Scurvy*, but it would be another 40 years before British sailors became known as "Limeys" for their rations of citrus fruits.

Another kind of experimentation was conducted on people who were classified as "monsters," including giants, conjoined twins, and intersexed people, then known as "hermaphrodites." The latter endured invasive and painful medical examinations far surpassing the norms of the time, when doctors barely touched their patients. In his report to the Royal Society, the physician James Douglas (1675–1742) described his examination of "hermaphrodite" Constantia Boon in 1715 with terms usually employed for dissection: he stated that he could not probe more deeply because the subject was still alive.

Vivisection in the Garden: Stephen Hales and Satire

Although Harvey had employed quantitative arguments, research on animals in the seventeenth century was largely qualitative. Mechanical models for animal form and function were not necessarily mathematical models. Boyle and Hooke did not attempt to measure the amount of air they pumped out of their air pump. Fifty years later, the English clergyman Stephen Hales (1677–1761) continued their experimental program of applying the principles of physics to the exploration of life. But the laws of motion Hales employed were those Newton had elaborated, the circulatory hydraulics he explored were analogous to the movements of the planets in the heavens, and numbers and measurement were at the center of his work. Hales wrote, "Since we are assured that the all-wise Creator has observed the most exact proportions, *of number, weight, and measure,* in the make of all things; the most likely way therefore, to get any insight into the nature of those parts of the creation, which come within our observation, must in all reason be to number, weigh and measure."[6]

Hales began experimenting on animals while a student at Cambridge University in the early eighteenth century. He and his friend William Stukeley (1687–1765) dissected and vivisected animals in Stukeley's college rooms. Stukeley was especially good at catching some of the many stray cats and dogs in Cambridge for their use. Hales went on to become a Church of England clergyman, but he continued to experiment. While at Cambridge, Hales attached a tube to a dog's artery and noted the rise and fall of blood at systole and diastole. This inaugurated his research on blood pressure, most of which took place in the garden of Hales's home in the village of Teddington, near London, where he became vicar in 1709. He continued to use dogs but performed his research mainly on horses because their size allowed Hales to easily measure their blood vessels and such little-known phenomena as blood pressure and the force of the heart. Hales attached a flexible tube to an opened artery and connected that tube to a long glass tube that acted as a pressure gauge. With each heartbeat the blood rose and fell in the glass tube; Hales measured the height to which the blood rose, noting that the pulse rate and the output of the heart varied with stress and exertion. He also compared the blood pressure, heart rate, and cardiac output of large animals with those of small animals, finding that the blood pressure was higher in larger animals, while the output of the heart was proportionately greater in smaller ones.

Hales was not a physician, and his work did not have any immediate medical application—nor was it meant to—yet it eventually led to the development of instruments for measuring blood pressure, as well as to an understanding of its medical significance. Hales's interest was hydraulics, and he employed very similar techniques in his study of plants, fixing tiny tubes to the veins of plants in order to measure the force of the flow of sap. Hales turned to plants because, he said, of the "disagreeableness" of experimenting on animals. He eventually returned to animal research.

Hales's work with horses could hardly have been a secret in a small

3.2 The Anatomy of the Horse

Horses were uncommon experimental animals. They were too big for most dissection rooms. In addition, they were valuable, even indispensable, for transport and farm labor. In *Gulliver's Travels*, Swift contrasted the Houyhnhnms, a race of intelligent, cultured horses, to the savage, filthy humans he called Yahoos. The first anatomical text that treated an animal in as much detail as a human was Carlo Ruini's 1598 *Anatomia del cavallo, infermità, e suoi rimedii* (Anatomy of the horse, its diseases, and its remedies). Ruini closely followed the format of human anatomical texts. Veterinary medicine, which developed as a discipline in the eighteenth century, with the first schools in France at Lyon and Alfort, grew out of both human medicine and farriery, which was defined as horseshoeing but which covered all aspects of horse care and health.

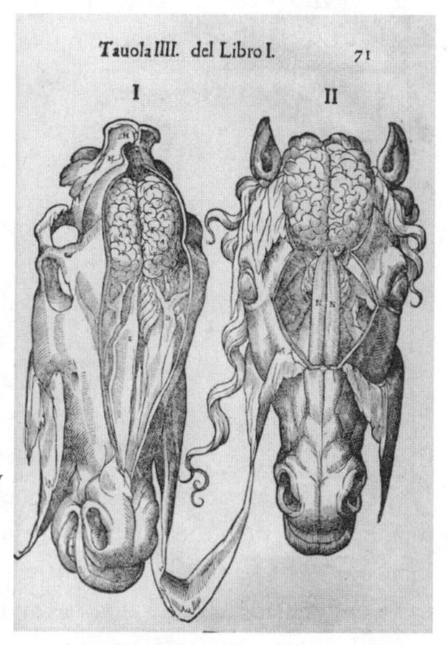

Carlo Ruini, *Anatomia del cavallo*, 1599. Bibliothèque interuniversitaire de santé, Paris, Licence ouverte

village, and it is likely he enlisted the help of his parishioners to tie down the horses he used. There is a Monty Python grotesquerie in the image of this shy bachelor clergyman (he married at 43, but his wife died soon thereafter) cutting the throats of horses in the back garden, and Hales indeed became a target for critics and satirists. While Boyle and Hooke carried out their experiments with little outside criticism, Hales was not as fortunate. Although the English public gradually accepted experimentation on humans in the form of smallpox inoculation, some of its members were less tolerant of experimentation on animals.

In coffeehouses and taverns, science had become a popular topic of discussion. In the 1690s, condensed versions of the *Philosophical Transactions* of the Royal Society of London and other scientific works—in English, not Latin—appeared. All over Europe and in the American colonies, scientific literature was readily available, either in scientific journals or in new popular magazines and newspapers. Because science was only beginning to develop a specialized vocabulary, most educated people could understand these accounts. Science was not yet a profession, and an interested amateur like Hales was as much the norm as a paid experimenter like Hooke. Instruments such as air pumps often formed part of aristocratic "cabinets" of curiosities.

In this age of satire, science was often a target. In the 1670s, as we saw in chapter 2, Thomas Shadwell's play *The Virtuoso* took aim at Royal Society experiments through its portrayal of Sir Nicholas Gimcrack. In 1726, the Irish writer Jonathan Swift (1667–1745) published *Gulliver's Travels.* On the flying island of Laputa, Gulliver found intellectuals so unworldly that they required "flappers" to slap them in the face periodically and snap them from their perpetual reverie. Laputan exiles founded the Academy of Projectors of Lagado, which bore some similarity to the Royal Society. Its members occupied their days with ridiculous and often repulsive experiments, many on animals. In 1711, the popular magazine the *Tatler* referred to idle physicians who cut up dogs and put cats in an air pump for amusement. The same satirist referred to Colombo's 1550s experiment with a pregnant dog as if it were a current event.

Hales's work on animals was widely known, especially following the 1733 publication of *Haemastaticks*. Just two years earlier, an article in the popular *Gentlemen's Magazine* had advocated vegetarianism. Now Hales's neighbor, the poet Alexander Pope (1688–1744), referred to him as having hands "imbrued with blood." Pope's essay "Against Barbarity to Animals" had appeared several years earlier, and his love of dogs, some of whom appeared in his poems, was well known. Another poet referred to

> Green *Teddington*'s serene retreat,
> For philosophic studies meet,
> Where the good Pastor, *Stephen Hales*,
> Weigh'd moisture in a pair of scales:
> To ling'ring death put mares and dogs,
> And stripp'd the skin from living frogs
> (Nature he loved, her works intent
> To *search*, and, sometimes, to torment!)[7]

Standards of acceptable behavior continued to change. In the eighteenth century, the new urban culture of coffeehouses and theaters demanded new modes of behavior centered on the notion of politeness. Shadwell portrayed Sir Nicholas Gimcrack as absurd but not especially cruel, but the literary critic Samuel Johnson (1709–1784) referred in 1758 to "wretches, whose lives are only varied by varieties of cruelty; whose favourite amusement is to nail dogs to tables and open them alive."[8] Johnson also reasserted the old Christian argument that physicians who performed animal experiments were thereby made cruel. Cruelty to animals increasingly became one aspect of now unacceptable behaviors.

Swift's account of the Houyhnhnms and the Yahoos questioned what had previously been assumed, that humans were innately superior to other animals. While his question remained unanswered, that he could ask it, even satirically, indicated that a new attitude was evolving.

Vitalism and Mechanism

Although Hales was a mechanist, by the 1730s the mechanical philosophy was being challenged in biological explanation. Even though he used mechanical imagery, Harvey did not believe that the body operated strictly according to mechanical principles, asserting that a vital principle separated life from nonlife. Scientists dissatisfied with mechanical explanations now revived these *vitalist* ideas. Yet there was not a dichotomy between mechanists and vitalists, and there were many gradations and syntheses between the two.

The basic principle of mechanism was that the organism could be explained solely according to laws of physics, that is, in terms of matter and motion. Vitalism's basic premise was that the organism was essentially different from inorganic nature. This difference was usually expressed in terms of a *vital force* not reducible to the laws of physics. In the eighteenth century and into the nineteenth, the debate between mechanism and vitalism encompassed a broad investigation into the workings of the animal and the human body, including generation, muscular motion, metabolism, and digestion. Both mechanists and vitalists experimented on animals to prove their assertions; indeed, because vitalists firmly believed that life and nonlife were essentially different, some were perhaps even more likely to experiment on live animals. Cartesian mechanical philosophy had ceased to be—if it ever had been—the main motivation for animal research.

The Swiss Abraham Trembley (1710–1784) made an important argument in favor of vitalism with his 1742 account of the freshwater hydra, a

3.3 The Best Experimental Subject

Researchers have experimented on themselves for centuries. Who could be a better research subject? In the early seventeenth century, the Italian physician Santorio Santorio (1561–1636) spent most of his days for thirty years on a large balance, or scale, which allowed him to measure his intake of food and drink and his bodily discharges. He concluded that the body loses each day a quantity of fluid that he referred to as "insensible perspiration." This was an important step toward understanding metabolism. A century later, another Italian, Lazzaro Spallanzani (1729–1799), investigated the digestive process, particularly the role of the gastric fluid in digestion. An earlier researcher had made tame birds swallow small, perforated tubes filled with food in order to investigate this process and the differing rates of digestion of different foods. Spallanzani performed similar experiments using himself as an experimental subject. He swallowed cloth bags and wooden tubes with various foods inside, which he later vomited up and studied. He also swallowed sponges to retrieve samples of stomach fluid. With these samples he conducted in vitro experiments, which showed that contrary to contemporary theories, digestion was primarily a chemical process accelerated by heat.

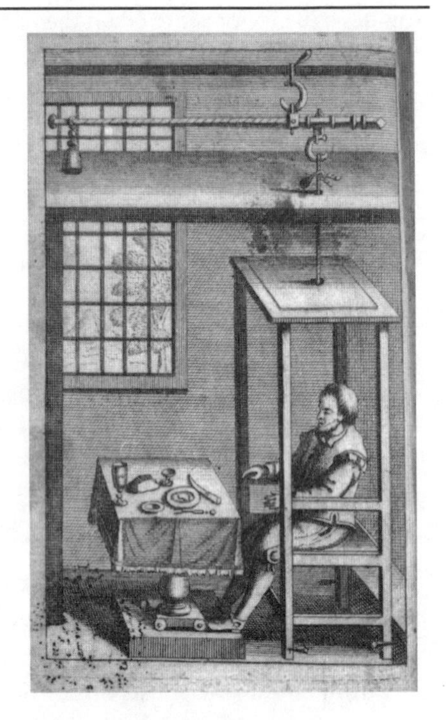

Santorio in his chair, *La médecine statique de Santorius*, 1722. Bibliothèque interuniversitaire de santé, Paris, Licence ouverte

small and primitive animal. This so-called polyp possessed the power of regeneration. When Trembley cut it up, each piece regenerated a new polyp; life, he concluded, is inherent in living tissue and cannot be fully explained by the arrangement of its parts. Others noted that frogs and newts could also regenerate parts. Yet five years later the Frenchman Julien Offray de la Mettrie (1709–1751) issued his uncompromising materialist manifesto *Machine Man*, which denied any special vital or mental properties even to humans.

Trembley's account, rather than La Mettrie's, inspired research on the fundamental question of the nature of life. Can the basic parts of the organism be defined as living, or is only the organized being or organism so defined? The most significant contribution toward resolving this question was made by another Swiss, the physician Albrecht von Haller (1708–1777). Haller, a professor of medicine at Göttingen in Germany, was one of the best-known physicians of his day and a strong advocate of smallpox inoculation. He performed a lengthy series of experiments on animals in the early 1750s on the quality of irritability in living tissues, defined as the ability of living tissue to respond to stimuli, usually by contraction. This quality was first noticed by Francis Glisson (1597–1677), who thought all body fibers were irritable. After Glisson's work, discussion centered on mechanical as opposed to vital explanations of irritability: was there a way to explain irritability, and more generally muscular motion, in mechanical terms? Hales had attempted to find a hydraulic explanation but had concluded that blood pressure had little to do with muscular motion. Haller sought instead to give a purely experimental definition, following Isaac Newton, who had claimed he simply described phenomena without assigning causes.

To Haller, irritability was the ability of a muscle fiber to contract upon stimulation. It was an unconscious response of the organism, dependent not on the nervous system but on a quality inherent in muscle tissue. He classified this phenomenon on a scale of highly irritable (contracting with slight stimulation) to slightly irritable (contracting with heavy stimulation). He also identified a second independent property that he called sensibility. Sensibility was a conscious response of the organism to stimuli and therefore a property of tissues that have nerves. Pain was an example of sensibility. Haller reached these conclusions, first published in 1752, as a result of experiments on nearly 200 animals in which he stimulated various parts of the body and recorded the response. The sources of stimuli ranged from touching to heat, cutting, and acids. Haller argued that experiments, being repeatable, led the way to truth. He emphasized the importance of comparative anatomy in establishing the functions of the animal body: only by comparing many species could one ascertain which structures and functions were common to all and which were specific to an individual. Haller was philosophically a vitalist, and irritability was to him an example of a vital function that could not be reduced to mechanics.

From the middle of the eighteenth century on, another school of vitalist thought emerged at the medical school of Montpellier, France. Much like the ancient empiricists, this school rejected experimentation because it interrupted the flow and spontaneity of life. They argued that life was essentially different from death and could not be determined by the fixed laws of physics. Haller did not attempt to find a physical, material explanation for irritability. But he believed that if, as he had concluded, experimentation was a valid method because experiments were repeatable, then nature must be determined by certain fixed laws (a point of view later known as *determinism*). He did not believe life was spontaneous in the way the Montpellier vitalists argued.

"But, Can they *suffer*?"

Haller justified his animal experiments as follows: "I have examined several different ways, a hundred and ninety animals, a species of cruelty for which I felt such a reluctance, as could only be overcome by the desire of contributing to the benefit of mankind, and excused by that motive which induces persons of the most humane temper, to eat every day the flesh of harmless animals without any scruple."[9] That Haller felt he needed any justification at all signaled the continuing debate about the morality of animal experimentation. In 1740, the *Gentleman's Magazine* published a satirical poem about air pump experiments. Five years later, a German literary journal published a defense of animal experimentation by Christlob Mylius (1722–1754), a medical student, who argued, much as Haller had, that the benefits to humankind outweighed the costs to animals.[10]

In the notable year 1789, Jeremy Bentham (1748–1832), English philosopher of law and political institutions, enunciated a new view of the relationship between animals and humans. Bentham published the principles of his *utilitarian* political philosophy in his *Introduction to the Principles of Morals and Legislation*. He defined *utility* as the property of producing good, pleasure, or happiness in an individual. The goal of government and society—both of legislation and of ethics—should be utilitarian, actively prohibiting pain or unhappiness. According to Bentham, humans were not the only beings susceptible of happiness and therefore within the ethical realm: animals also qualified.

Although legal theory had relegated animals to the status of things, Bentham argued that animals could be considered interested parties under utilitarianism because they were capable of happiness, and he enumerated

William Hogarth, *The Four Stages of Cruelty*, plate 4, *The Reward of Cruelty*. Metropolitan Museum of Art

3.4 The Four Stages of Cruelty

In his series of engravings *The Four Stages of Cruelty*, the English artist William Hogarth (1697–1764) traced the criminal career of a coachman named Tom Nero, which began with torturing dogs, went on to the beating of horses, and ended with murder. The German philosopher Immanuel Kant (1724–1804) had Hogarth's engravings in front of him in 1780 when he restated the Christian position on animals. Cruelty to animals, said Kant, would damage human moral sensibility and therefore was to be avoided, but humans had no moral obligations toward animals.

James Gillray, *The Cow-Pock*, engraving, 1802. Library of Congress

3.5 The Cowpox Vaccine

Edward Jenner's smallpox vaccine, based on cowpox, generated a huge amount of publicity. James Gillray (1757–1815), a well-known caricaturist, contributed this view in 1802 of Jenner injecting his clients with cowpox after administering an "opening mixture" in preparation.

their interests. Humans could, he said, justly kill and eat animals: "The death they suffer in our hands commonly is, and always may be, a speedier, and by that means a less painful one, than that which would await them in the inevitable course of nature." But he denied that humans could justifiably cause animals to suffer, comparing the status of animals to that of slaves (Bentham was active in the burgeoning antislavery movement in England). "The day *may* come," he added, "when the rest of the animal creation may acquire those rights which never could have been withholden from them but by the hand of tyranny." The number of legs might be as irrelevant as the color of skin in assigning rights to an individual, and the Cartesian criteria of cognitive ability and speech would be equally irrelevant. The question, Bentham eloquently stated, "is not, Can they *reason*? nor, Can they *talk*? but, Can they *suffer*?"[11] Although Bentham's argument was overlooked in the nineteenth century, it ultimately strongly influenced twentieth-century discussions about the rights of animals. He did not mention animal experimentation, however, and the argument of utility came to be used on both sides of that debate.

Jenner and Vaccination

At just the time when Bentham was writing, inoculation against smallpox took an unexpected turn that emphasized the close relationship between certain animal and human diseases. Inoculation had continued throughout the eighteenth century, and historians generally agree that it had an impact on reducing the number of deaths from smallpox—although there is much disagreement about the size and significance of that impact. But inoculation, as we have seen, had its drawbacks. Although the Suttons claimed that their method of inoculation produced only a single pustule on the patient, no one understood how infection occurred, and inoculation could be dangerous to the patient and to those around him, who might contract smallpox even from that single pustule.

Inoculation against smallpox originated with folk practice, and dairymen and milkmaids had long known that if they contracted a bovine disease known as cowpox they were very unlikely to contract smallpox. Cowpox was common among cows, and ulcerated teats were one of its signs. If a milkmaid had a small cut on her hand, she could easily contract the disease when she milked an infected cow. Cowpox was usually a mild disease in humans, manifesting itself in a few pustules, often on the hands and arms, and about a week of feeling unwell. Edward Jenner (1749–1823), an experienced inoculator, was interested in cowpox and its causes and the interrelations among cows, humans, and horses. A physician in Gloucestershire, an English county known for its dairies and cheese, Jenner had studied in London with John Hunter (1728–1793), one of the most important naturalists and comparative anatomists of his day. Jenner believed that a horse disease known as "the Grease" could manifest itself as cowpox in cows.

During the 1780s and 1790s, Jenner collected case histories of patients who had contracted cowpox in the past and then were found to be resistant to smallpox, either natural or inoculated. He documented 23 cases in his 1798 *Inquiry into the Causes and Effects of the Variolae Vaccinae*. Jenner began this work not with the cow but with the horse, claiming that "the Grease" was capable of generating a smallpox-like disease in humans that he thought it "highly probable" might be "the source of that disease." Jenner argued that "incautious" milkers could transmit this horse disease to cows, where it manifested itself as cowpox, which could then infect humans. "What renders the Cow-pox virus so extremely singular," wrote Jenner, "is, that the person who has been thus affected is for ever secure from the infection of the Small Pox."[12]

In most of the cases Jenner narrated, the infection with cowpox had occurred many years earlier, and immunity was only demonstrated by later exposure or inoculation. Based on this evidence, Jenner determined to test his theory. Case 17 narrates Jenner's first "vaccination" (from *vaccinia*, "cowpox," via *vacca*, Latin for "cow"). On 14 May 1796 Jenner inoculated an 8-year-old boy (unnamed in the book but identified as James Phipps). Jenner used material he had taken from the arm of Sarah Nelmes, a milkmaid who was infected with what Jenner believed to be cowpox. James Phipps received the cowpox matter in his arm in the same way others were inoculated with smallpox matter. He felt some discomfort over the course of the next ten days but recovered. In July 1796 Jenner variolated (that is, inoculated with smallpox, or *variola*) young Phipps in both arms, and nothing happened. Phipps did not get smallpox. Jenner repeated this several months later, and indeed several more times over the next few years. While Jenner's experiment appeared to have been successful, these repeated variolations may also have built up Phipps's immunity to smallpox. Jenner also injected another small boy, John Baker, with matter from a farmhand who had been infected with "the Grease." Baker experienced cowpox-like symptoms, but Jenner was unable to inoculate the boy with smallpox to test his immunity because he fell ill with another fever in a workhouse, a public institution for the very poor.

Jenner extended his experiments, creating a chain of vaccinations from the original subject, who contracted cowpox from a cow. He then used material from that subject to vaccinate a second, and so on through several other subjects, most of them children. This "arm-to-arm" method, also used in variolation, became standard practice in nineteenth-century vaccination. But Jenner did not then test (or *challenge*, in modern experimental parlance) all of those vaccinated by variolation. Nonetheless, Jenner was convinced that vaccination was an effective and safe alternative to variolation. By referring to *vaccinia* as *variolae vaccinae*, that is, "cow smallpox," Jenner asserted a necessary relationship between the two diseases.

Others tested the effectiveness of cowpox vaccination, especially Dr. William Woodville of the London Smallpox and Inoculation Hospital, but his results only confused the issue more, since it was likely that Woodville's vaccine—which passed through several individuals—was contaminated with the smallpox virus. By employing human rather than animal sources of cowpox, Woodville and others distanced the vaccine from its rural and animal origins. Children continued to be incubators for the vaccine. When King Carlos IV of Spain decided in the early nineteenth century to provide vaccination throughout his empire, the royal surgeon, Don Francisco Xavier Balmis (1753–1819), took 24 orphans with him on his global journey to maintain supplies of the vaccine. Sixty years later, a Confederate surgeon employed slave children for the same purpose. The "vile bodies" of destitute and powerless children enabled vaccination to continue.

We now know that Jenner was essentially correct, although neither he nor his contemporaries understood why. He was correct about the relationship of cowpox to horses as well. The microbiologist Derrick Baxby (1940–2017) surmised in 1977 that "the Grease" was identical to horsepox and that horsepox, not cowpox, was the basis of Jenner's vaccine. Jenner himself performed what he called "equinating" as well as vaccinating, and recent genetic work has employed horsepox to develop a new smallpox vaccine.[13]

A full understanding of the action of vaccines in inducing immunity took more than a century to develop, following from Pasteur's work in the 1880s (see chapter 5). Jenner, like the early variolators, was taking a shot in the dark. Like theirs, his patients were his experimental subjects, and many of them were children. It cannot be denied that Jenner risked the lives of his patients. It also cannot be denied that naturally caused smallpox was declared definitively eradicated on earth in 1980. The connection between these two occurrences cannot be easily reduced to a cost-benefit analysis.

The controversy over smallpox inoculation in the eighteenth century is relevant to many issues surrounding experimentation on humans today, particularly consent and acceptable risk. Inoculation (variolation) helped more than it harmed, but no one, not even the physicians, had sufficient information to predict the outcome in a particular case. Jenner's vaccination minimized some risks but introduced others. While this large-scale human experiment went on, however, experiments on animals were scrutinized more and more closely, and critics expressed increasing concern about cruelty to animals and its impact on humans. All of these issues would only intensify in the nineteenth century.

4 Cruelty and Kindness

Early in 1825—two centuries after Harvey's discovery of the circulation of the blood—Richard Martin (1754–1834), a member of the British Parliament, spoke to the House of Commons about his bill to abolish bearbaiting and other animal sports. He concluded: "There was a Frenchman by the name of Magendie, whom [Martin] considered a disgrace to society. In the course of last year, this man, at one of the anatomical theatres, exhibited a series of experiments so atrocious as almost to shock belief."[1] This Magendie had nailed "a lady's greyhound" to a table and then proceeded to perform horrific operations upon its face and body, which Martin described in graphic detail amid cries of "Shame" and "Hear, hear!" Martin's bill failed, however, and doubts emerged about the story of the greyhound, some saying it had not happened at all, others that the dog had not been a greyhound but a spaniel. Nonetheless, the story established Magendie's negative reputation in Britain and focused antivivisection sentiment across the English Channel.

Who was Magendie, and why did he arouse such furor? François Magendie (1783–1855) was one of a group of French scientists who established experimental physiology as a key science. This group coined the very term *physiology*, aiming to discover how organisms worked, as well as the interdependent functions of a living body. Anatomy, chemistry, and physics all played roles in this investigation, but vivisection was at its center. Moreover, they referred to no single philosophical system. On the issue of vitalism versus mechanism they remained neutral.

Harvey, Haller, and many others had experimented on animals. What was different about Magendie and his circle? Earlier experimenters had sought to answer specific questions about animal (and human) function, questions posed by theories that had not been determined by experimental means. They had not established a tradition of systematic experimental research, in which one experiment led to another. Harvey's followers, for example, considered several questions raised by circulation but did not link these together systematically to overturn traditional ideas. After Magendie, experimentation was, and continues to be, the distinguishing feature of research in physiology, no matter how physiological theories changed. Experimentation became the rule.

Magendie benefited from his predecessors. Events of late-eighteenth-century France also shaped his outlook. During the eighteenth century, medical schools in Leiden and Edinburgh moved away from texts and toward hands-on training, including observation of patients and dissection of cadavers. Surgery, the manual side of medicine, began to merge with the theoretical medicine of physicians, and surgeons' skills became more valued. Veterinary medicine also contributed to this reformation of medical theory and practice. New French veterinary schools in the 1760s at Lyon and Alfort established a model of professional education, and medical reformers assumed the unity of human and animal medicine. The availability of animals at the veterinary schools made them centers for experimentation, and the demands of cavalry warfare in the Napoleonic era only increased their value to the state. Many researchers trained at Lyon or Alfort.

During the French Revolution, which began in 1789, medical and surgical training focused on the hospitals. Revolutionary ideology discounted the past, emphasizing empirical knowledge and analysis rather than theorizing in medicine and the life sciences. The resulting French clinical school of the first third of the nineteenth century made Paris the acknowledged center of medical authority. While physiologists sometimes complained that medical reliance on clinical observation and autopsy ran counter to their own interest in experimentation, medicine provided physiologists with necessary skills and with interesting problems to study.

Magendie trained as a surgeon and began his research career at Alfort. His cohort took inspiration from Xavier Bichat (1771–1802), a surgeon and anatomist. In his *Discourse on the Study of Physiology* (1798), Bichat set forth the criteria for a new science. While mindful of the uncertainties involved in experiments on live animals, Bichat combined autopsy, anatomy, and vivisection into a powerful research program. Because vital forces were inherently unstable, animal experimentation required strict rules and careful procedures. Bichat established general rules for experimentation that became the basis of the new science of physiology:

- Compare the experimental animal with a normal one used as a control.
- Eliminate "accidental interferences"; that is, control the experimental environment as much as possible.
- Repeat the procedure to confirm that the same result is obtained.
- Carefully examine the state of the animal before and during the experiment.

Bichat's tissue doctrine stated that the organism, or even the organ, was not the fundamental unit of analysis in the body; simpler organic elements existed, which he identified as twenty-one different tissues. In the same way that chemistry in the 1790s had become a science of elementary bodies with the concept of the chemical element, so too anatomy, said Bichat, would be a science of elementary tissues, the study of simple organic "elements" and the structures they formed. The principle of life was at the level of the tissues. Bichat's tissue doctrine had two effects on research in the nineteenth century. First, it stimulated the search for even more fundamental levels of analysis than tissues, leading to cell theory in the 1830s and 1840s. Second, it established the experimental method as the most fruitful method for physiological research by confirming that organic function was uniquely biological. Bichat's definition of life as "the totality of functions which resists death" bypassed the vitalism/mechanism question and focused research on the organism.

Bichat believed that all tissues possessed the principle of life and therefore possessed some degree of sensitivity, feeling, and movement. Magendie and his circle disagreed with this view but retained Bichat's experimental focus. Their experimental work rejected Bichat's emphasis on the body as an organic whole and concentrated on isolated systems. Julien Legallois (1772–1814) argued that life centered in the spinal cord. He separated a nerve from its connection to the spinal cord, thereby depriving the affected part of sensation and motion. In addition, even parts that Haller had considered merely "irritable" had in fact some connection to the spinal cord, although Legallois could not entirely explain why the heart continued to beat in an animal whose spinal cord had been destroyed.

Legallois wrote, "Experiments on living animals are among the greatest lights of physiology. There is an infinity between the dead animal and the most feebly living animal."[2] This statement might serve as Magendie's motto.

Among the scientists of his era, Magendie remained the most devoted to experimentation. Viewing himself as the Newton of physiology, he detailed its principles in his new journal of experimental physiology, the *Journal de physiologie expérimentale*, aiming to make physiology as certain a science as Newtonian physics. Physiology, he believed, had been damaged by the speculative tendencies of vitalism, and like Newton, he claimed that he would merely observe and not draw conclusions about causes. Any function, Magendie said, was the sum of actions of several organs and did not depend on any vital property. The fundamental distinction was simply between living and nonliving, and the physical or chemical causes of so-called vital phenomena were not yet known. Experimentation established the anatomical organization of functions and their mode of action, revealing the laws of vital phenomena. Dissection of the dead could not do this.

Like Legallois, Magendie focused on the nervous system. His first experiments tested the action of a poison on dogs and other animals. He introduced the poison on a sliver of wood into the thigh of the animal, who shortly thereafter began to experience symptoms of poisoning, including paralysis. Magendie concluded that circulation carried the poison to the spinal cord, a conclusion he tested by introducing the poison into other parts of the body. Injection into parts well supplied with blood vessels led to rapid absorption; in other parts of the body it acted more slowly, and it acted even more slowly when ingested with food. Direct injection into the arteries produced rapid action. Experiments to separate the spinal cord from the brain and to destroy part or all of the spinal cord showed that the cord, as he had claimed, was the seat of action for the poison.

These experiments demonstrate Magendie's techniques and philosophy. He attacked a single question from many angles, testing and retesting systematically, never satisfied with obvious conclusions. Even facts that were apparently well known required new tests, new experiments. He viewed the organism as a series of individual systems, localizing feeling or sensibility in the nervous system and not in individual organs. In another experiment on the action of an emetic drug, he showed that it acted via the nervous system, not by the direct action of the stomach alone. In repeating experiments and testing alternate explanations he used many animals, and many kinds of animals. Many of these experiments caused pain.

Although Magendie occasionally used opium to anesthetize his experimental animals, his experiments on the spinal cord and the roots of the nervous system required that the animal be fully alert. In the early 1820s, a detailed and painstaking series of experiments on puppies and later on other animals demonstrated that each nerve had two roots (known as posterior and anterior, from their position) on the spinal cord; the posterior roots controlled sensibility, while the anterior roots controlled motor skills. In other words, when he cut a nerve at the posterior root, a dog could walk but not feel, while when he cut a nerve at the anterior root, it could feel but not walk. The organism could not feel; only the nervous system could. Life resided in the brain and the spinal cord, and the brain mediated feelings and emotions. Not since Galen had such systematic analysis of the nervous system been undertaken, with the extensive vivisection this entailed. Like Galen, Magendie often appeared to be cruel and unfeeling toward his experimental animals.

The Beginnings of Antivivisection in Britain

Richard Martin was one of many Britons repelled by Magendie's experiments. A British visitor to France in 1817 commented disapprovingly on the "mania for vivisections" of Magendie and his circle, but the Frenchman's experiments on the nervous system aroused much scientific interest. Magendie's research may have been inspired by a demonstration given at Alfort in the early 1820s by an assistant to the English researcher Charles Bell (1774–1842), and the differentiation of nerve functions is known as the Bell-Magendie Law. When Magendie went to London in 1824, therefore, he was well known among scientists, and the differences between his style of research and Bell's highlighted broader national attitudes toward animals. Where Bell dissected cadavers and dead animals, supplemented by vivisection (often on animals previously "stunned," which limited their ability to display sensibility), Magendie vivisected without hesitation. According to Richard Martin, Magendie said "Soyez tranquille" (Be still) to a restless dog during his London demonstrations, adding jokingly, "Il serait plus tranquille s'il entendait français" (he would be calmer if he understood French). The British press repeated this comment as an example of his heartlessness.[3]

How had these different attitudes arisen? Why was Magendie lauded in France, while he and the British researcher Marshall Hall (1790–1857) were vilified in Britain? One answer is that vivisection provided a touchstone for a multitude of ideas and emotions in an age of rapid and turbulent change in politics, culture, technology, and society, especially in Britain. The 1820s witnessed the disillusionment of the Romantics, whose hopes had been kindled by the revolutionary ideals of 1789. The political radicals Mary Wollstonecraft and William Godwin believed that science and reason could remake the world, but in *Frankenstein* (1818) their daughter, Mary Shelley (1797–1851), vividly refuted their claim: science and reason, pressed to their limits, led to horror, despair, and death. Victor Frankenstein built his creature of body parts from stolen corpses but did not then assume responsibility for his creation. His cruelty toward the creature ultimately led to his own death. Mary Shelley called Frankenstein "the modern Prometheus," referring to the figure in Greek mythology who stole fire from the gods and was punished for his hubris, for his overweening pride.

France remained overwhelmingly rural in the 1820s, but industrialization had already led to widespread urbanization in Britain. Romantic poetry in Britain idealized the countryside, both wild places and pastoral landscapes of farms and fields that were rapidly disappearing. Alongside the old social order with its clear boundaries of rank and station arose a new society of industrial entrepreneurs and restless workers who demanded a share of political power. The Great Reform Bill of 1832, while hardly radical, gave parliamentary representation to the new urban metropolises and extended voting rights for the first time in four centuries. Meanwhile, evangelical clergy pondered the morality of slavery, which was finally abolished in 1838. Traditional structures of society seemed to be collapsing along with traditional values. New urbanites, as well as old gentry, longed for a return to an ideal rural past of deference and paternal kindness.

Domesticated animals, above all dogs, provided the new urbanite with a connection to the rural past. Petkeeping, long an aristocratic privilege, became established among the middle classes at the end of the eighteenth century. The proliferation of dog (and cat) breeds in the nineteenth century helped distinguish middle-class pets from working animals and from the mutts of the lower classes. In addition, the fabled loyalty and faithfulness of dogs stood in contrast to the instability of modern life.

In 1800, a bill in the House of Commons opposing the "savage custom of bull-baiting" was greeted with derision and defeat, and Mary Wollstonecraft's 1792 *Vindication of the Rights of Women* was parodied by a *Vindication of the Rights of Brutes*—what could be more absurd? But in 1822, Richard Martin successfully guided his bill forbidding cruelty to farm animals (later amended to include dogs and cats) through the House of Commons.

William Austin, *The Anatomist Overtaken by the Watch . . . Carrying off Miss W——ts in a Hamper*, etching, 1773. National Library of Medicine

4.1 The Anatomy Act

Dr. Frankenstein stole bodies to make his monster. He "dabbled among the unhallowed damps of the grave, [and] tortured the living animal." In the illustration, the famous London anatomist William Hunter (1718–1783) is caught in the act of stealing a corpse for dissection. But a year after the second edition of *Frankenstein* appeared in 1831, grave robbing was no more in Britain. The Anatomy Act of 1832 gave the anatomists the bodies for which they clamored. Previously, anatomists and surgeons had had legal access only to the bodies of executed criminals. The "Murder Act" of 1752 had made dissection part of the punishment for murder. The growth in importance of anatomical studies made this supply of bodies inadequate for the purposes of the anatomists, who resorted to robbing graves. In Paris, Bichat was known for his skills in cemetery raiding. The case of Burke and Hare in Edinburgh in 1828 brought the issue to the fore, and arguments surrounding the anatomy bill played on the complicated relationship between human and animal bodies, life and death. The Anatomy Act stated that the bodies of the destitute who died in government-run workhouses and hospitals would be available for anatomists. From being a punishment for murder, dissection became a punishment for being poor.

■ Quote from Mary Shelley, *Frankenstein* (London, 1818), 90.

Martin was aided by the antislavery clergyman William Wilberforce, along with a mass of petitions, mostly from urban areas.

The defeat of Martin's bill in 1825, in the course of which Martin decried Magendie's barbarous science, was only a momentary setback. In 1835, Martin's Act of 1822 was amended to cover animal sports. In 1824, the first animal protection society, the Society for the Prevention of Cruelty to Animals, or SPCA, was founded in London. Queen Victoria's sponsorship later added *Royal* to the society's name. The RSPCA, whose membership was largely urban and upper class, focused on animal sports that Martin had sought to outlaw in 1825, such as bearbaiting. These were sports of the lower classes—the RSPCA did not address the upper-class practice of fox hunting—and Martin argued that banning these sports would help to civilize the naturally violent lower orders. To be humane was a sign of civilization, and following the old argument, which made Wilberforce and Martin natural allies, humans who were cruel to animals would also be cruel to their fellow humans.

Martin was a founding member of the SPCA, and these early animal protectionists opposed vivisection, although there were few examples of it in Britain. Steady attention from both the popular press and medical journals nonetheless kept the topic current. In 1829, an editorial in the *London Medical Gazette* recalled the popular outcry over Magendie's visit five years earlier in arguing in support of what became the 1832 Anatomy Act (see sidebar). A year earlier, the trial of Burke and Hare in Edinburgh, who murdered indigents and then sold their bodies to Dr. Robert Knox of the University of Edinburgh Medical School for dissection, had highlighted scientists' need for cadavers but also revealed heartless and unscrupulous anatomists, recalling once more their association with the hangman.

Foremost among British vivisectors was Marshall Hall, an Edinburgh-trained physician who kept a laboratory in his house while continuing a medical practice. In the 1820s he began to investigate the effects of blood loss. A few years earlier, James Blundell (1790–1878), a London obstetrician, had suggested blood transfusion to mitigate the consequences of uterine hemorrhage after birth and had experimented on dogs before transfusing several patients, some of whom survived. Hall also based his first publication on patient observations, but by the late 1820s he too was experimenting on dogs. Since bloodletting continued to be a popular therapy, this research had considerable practical significance.

Perhaps inspired by Magendie, Hall by 1830 was studying the interaction between the nervous system and the circulation of the blood. In the following year, Hall published *A Critical and Experimental Essay on the Circulation*, which described experiments on fish and frogs that included crushing the brain and spinal cord as well as other parts of the body to observe the effects on circulation. Mindful of antivivisection criticisms, Hall called for the formation of a "society for physiological research," which would regulate animal experimentation, since, he said, "every experiment . . . is necessarily attended by pain or suffering of a bodily or mental kind."[4] He proposed five guidelines for the performance of animal experimentation:

1. Experiments should be absolutely necessary, and all alternatives should be explored.
2. Experiments should have clear and attainable objectives.
3. Experimenters should avoid unwarranted repetition and should therefore be aware of previous experiments.
4. The least possible pain should be inflicted. (In this age before anesthesia, Hall recommended using "lower" animals such as frogs and fish or newly dead animals.)
5. Experiments should have witnesses to certify results and lessen the need for repetition.

The society never materialized. Hall, who continued to experiment, became a focus for antivivisectionists in the 1830s and 1840s. In 1847 Hall's reply to an extended critique of his work led to a series of publications, including an antivivisection pamphlet from the RSPCA. Hall continued to work on his own, with no official position or support, and when he died in 1857 he was still one of very few experimental physiologists in Britain.

The Introduction of Anesthesia

In 1846, amid the debates between Hall and his opponents, the first operation that successfully employed ether anesthesia was performed in Boston, Massachusetts. The introduction of anesthesia ultimately had a profound impact on perceptions of pain, changing the relationship between doctor and patient, as well as between experimenter and animal. Its use also revealed strongly held attitudes about the varying capacity for pain among humans and animals.

Pain has meant different things in different cultures, and as we have seen in Descartes's work, pain and suffering were not necessarily the same thing. Pain was a constant companion in premodern society. Today, for everyday pain such as headaches we can swallow a pill and expect relief. Before the twentieth century there was no such certainty. Christian theology viewed earthly suffering as a necessary prelude to heavenly bliss. Because pain was subjective, its relief was not the focus of healing practices, and many premodern cures such as bleeding and cauterization caused more pain. Physicians believed that pain, like fever, was nature's reaction to a bodily crisis and that its suppression could be harmful. The pain of surgery, therefore, was necessary to the healing process, and some claimed that more painful operations had greater success than less painful ones. Recalling Descartes, some physicians argued that the ability to feel pain was what made us truly human.

Opium, in the form of liquid laudanum, was available, as was alcohol, but neither of these was entirely satisfactory as an anesthetic agent: dosages were unpredictable, and the side effects undesirable. The quantity of either drug needed to produce total stupefaction was so great as to be dangerous, and experimental techniques for evaluating the relationship between a lethal dose and an effective dose had not yet been developed. The English writer Frances Burney (1752–1840) took a glass of sherry before she endured a mastectomy in 1811; her horrifying account of the operation in a letter to her sister Esther reveals that the alcohol had little effect on either her pain or her terror.[5]

Morphine, the active ingredient in opium, was isolated at the beginning of the nineteenth century. Several experiments on animals, especially dogs, followed to determine its mode of action: was it a stimulant, a depressant, or a poison? By the late 1820s, hospitals employed morphine both for pain relief and for many other ailments, and its addictive qualities became clear. New pain-relieving drugs, together with the new status afforded the sensitive character, meant that pain relief increased in importance to medical care over the course of the nineteenth century.

The chemistry of gases began to be known at the end of the eighteenth century with the isolation of oxygen and the discovery of several new gaseous compounds. Among these was nitrous oxide. The chemist Humphry Davy (1778–1829) tested nitrous oxide on himself and his friends in the early nineteenth century. He noted both its anesthetic effects and the pleasurable rush he called "the thrilling." But nitrous oxide did not come into use as an anesthetic until many years later, and then as a supplement to more potent agents. Anesthesia proved to be particularly fertile territory for self-experimentation. Many of the early developers of surgical anesthesia, including the dentists William T. G. Morton (1816–1868) and Horace Wells (1815–1848) and the physicians James Young Simpson (1811–1870) and Crawford Long (1815–1878), first tried the drugs on themselves. Morton tested ether on himself in his office. Wells experimented with nitrous oxide and later became addicted to chloroform, ending his life as a suicide. Long first encountered ether as a recreational drug in the early 1840s. Simpson, an Edinburgh obstetrician, tried several inhalants upon himself (including acetone and benzene) before settling on chloroform as the best anesthetic.

In contrast to opium or alcohol, ether and its near companions nitrous oxide and chloroform acted quickly. Nitrous oxide in low concentrations gave only an incomplete state of anesthesia, while chloroform was highly potent. When inhaled in the form of a vapor, ether put the subject into a state of unconsciousness resembling sleep, which painful stimuli did not disrupt. The topic of anesthesia divided the medical profession, and some physicians worried that the patient experienced pain but forgot it. Nonetheless, inhaled forms of anesthesia grew in popularity. James Young Simpson, in Edinburgh, employed chloroform for a difficult childbirth in 1847. Its use spread despite critics' warning that pain in childbirth was necessary to the process of giving birth, both because it stimulated contractions and because it provoked maternal love for the new child. (Chloroform relaxes the uterine muscles, thus diminishing the effectiveness of contractions.) Others declared that pain in childbirth was a necessary consequence of Eve's sin.

Experimenters, especially in France, used animals to discover how ether and other anesthetic gases worked: was loss of consciousness caused by the chemical or by lack of oxygen? How long could ether anesthesia be prolonged without ill effects? Cases of death by anesthetic asphyxia were not uncommon. Magendie objected to the use of ether on two grounds: because its mode of action was unknown, using it on patients amounted to human experimentation, which he could not countenance; and he found it unethical to operate on an unconscious person, who could not assess the surgeon's performance during the operation and was therefore at his mercy.

As with smallpox inoculation, the benefits of anesthesia seemed to outweigh its risks. However, medical men believed that the capacity for pain differed widely among humans according to "age, sex, and temperament,"

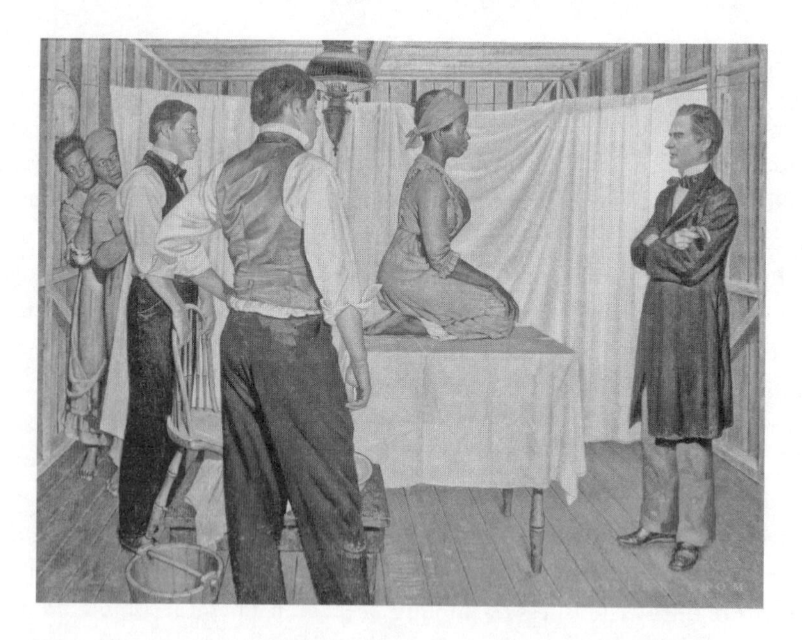

Robert Thom, from *Great Moments in Medicine*, ca. 1952. In the 1950s, the pharmaceutical company Parke-Davis commissioned a series of paintings suitable for doctors' offices. One painting in the series depicted Sims about to examine Anarcha. Courtesy of University of Michigan Art Museum

4.2 Slaves as Experimental Subjects

As we saw in chapter 3, John Quier's inoculation of slaves in eighteenth-century Jamaica walked a line between therapy and experimentation. In the antebellum US South, white physicians used black slaves as experimental subjects because they were available and legally invisible.

In Georgia in the 1820s and 1830s, Dr. Thomas Hamilton performed experiments on heat stroke on a slave named Fed. On several occasions, Fed sat in a pit surrounded by hot embers to test various remedies. Hamilton never published his results. In Alabama in the 1840s, Dr. James Marion Sims (1813–1883), later known as the "father of modern gynecology," experimented with surgical techniques in the repair of vesico-vaginal fistula (an opening between the bladder and the wall of the vagina, usually following childbirth) on three enslaved women, Lucy, Anarcha, and Betsey. Each woman experienced around thirty operations, which in this preanesthetic, preantiseptic age were both painful and risky. Sims finally found a successful technique and noticed that his white patients tolerated the pain of the operation less well than the enslaved women, perpetuating a myth of racial difference in pain perception. A statue of Sims in New York's Central Park was removed in 2018.

according to Magendie,[6] but also race and social class. Categorization by the ability to feel pain provided scientific affirmation of the ideal social order and reflected power relationships in Western society. Women, of weaker nerves, were generally believed to be more sensitive to pain than men, and the upper classes felt pain more acutely than the lower classes. Apart from temperamental differences, which tended to differentiate classes, the hard labor of the poor inured them to pain. Different races—which in nineteenth-century America included the Irish and other immigrant groups as well as Africans, Asians, and Native Americans—were also seen as feeling pain differently, whites being deemed the most sensitive. Children occupied an ambiguous status in this hierarchy of pain. Some physicians believed that children, like women, were more sensitive to pain. But others argued that young children, lacking reason, resembled animals in their relative insensitivity. Even today there is considerable argument about the capacity of newborns to feel pain, and until quite recently the use of anesthetic agents in such operations as circumcision was much debated.

Animals were at the bottom of this hierarchy, below the least sensitive human. Many believed that animals simply did not feel pain with the same intensity as humans, although, mirroring the human hierarchy, domestic animals were thought to be more sensitive than wild ones. Since operations on certain classes of humans continued to be performed without anesthesia until late in the nineteenth century, it is not surprising that the use of anesthesia on experimental animals was inconsistent. It was employed—as it sometimes was in surgery on humans—as much to keep the animal still as to relieve pain. Paralytic agents such as curare, for example, did not relieve pain. While it seems that the introduction of anesthesia would have eliminated a major objection to vivisection—that it was painful—in fact it raised more questions than it resolved. It freed the consciences of some scientists to pursue animal experiments to which they had previously objected on the grounds of cruelty and therefore led to an increase in invasive experiments. One antivivisectionist commented, "I am inclined to look upon anesthetics as the greatest curse to vivisectible animals."[7] It also forced antivivisectionists to examine the bases of their beliefs.

Claude Bernard and the Defense of Experimentation

When Magendie died in Paris in 1855, the French recognized him as one of the great scientists of the century. Unlike Hall, who had never won a university position, Magendie held a chair at the elite Collège de France, and his most promising pupil, Claude Bernard (1813–1878), succeeded him there. Bernard continued Magendie's research program and his campaign to establish experimental physiology as a distinct science. At Bernard's death, the methods of experimental physiology—centered on animal experimentation in the laboratory—had become the favored means of gaining knowledge about the human and the animal body. While Bernard made many significant discoveries, this section focuses on his public role as spokesman for experimental physiology, a role that he made explicit in his *Introduction to the Study of Experimental Medicine* (1865) and that made him a target for antivivisectionists. With Bernard, the laboratory rather than the clinic or the dissecting table became the primary site for learning about the body and its functions. It also became the place students came to learn about the body, a site for training as well as for research.

Bernard entered medical school in 1834 and began to work with Magendie five years later. Focusing first on digestion, he combined animal experimentation with chemical analysis to demonstrate the interconnections between the nervous system and individual organs in digestion and metabolism. He defined the scope of experimental physiology as encompassing not the nature of life but the experimental resolution of vital phenomena. The nature of life was unknowable, but the specific circumstances of a single phenomenon could be experimentally determined. Although living organisms were complex and changeable, their individual operations, such as respiration and digestion, followed laws and could be analyzed experimentally. Bernard presented his experimental results as an example of his philosophy of physiological determinism, which stated that experiments always produced replicable results and yielded identical outcomes.

Bernard believed that only experimental physiology could provide new information about the body, which could then be applied to human medicine. Although he had a medical degree, Bernard never practiced medicine and scorned its current practice. Its reliance on observation of sick patients and the anatomy of dead bodies gave no information about vital processes, which required the active intervention of vivisection. Nor could chemistry alone unlock the secrets of the body, although Bernard employed chemical analysis. His new science combined chemistry and analysis of tissues and cells with vivisection. Animals served as proxies for humans, and Bernard's goal, which ultimately eluded him, was to outline a general physiology for

4.3 Alexis St. Martin's Stomach

In 1822, a French-Canadian fur trapper in northern Michigan named Alexis St. Martin (1794–1880) received a gunshot wound in his side. His wound healed but did not close, leaving an opening into his stomach. This gastric fistula provided a window into the mysteries of the digestive system, and his doctor, army surgeon William Beaumont (1785–1853), decided to investigate these. St. Martin moved into Beaumont's home as a servant, and Beaumont began to poke bits of food into St. Martin's stomach. He also took samples of gastric fluid and frequently inserted a long thermometer deep into St. Martin's stomach, which was especially painful. Between 1825 and 1833 Beaumont performed hundreds of experiments on St. Martin, eventually publishing a book, *Experiments and Observations on the Gastric Juice, and the Physiology of Digestion* (1833).

St. Martin was not always a willing participant in these experiments, periodically leaving Beaumont and returning to Canada. He signed an employment contract with Beaumont in 1832 that bound him to the doctor for a year, in exchange for room and board and a sum of money. St. Martin returned home permanently at the end of that year and resisted further pleas for more experiments, although he lived in poverty.

Beaumont's experiments on St. Martin made both famous. Physicians and scientists praised Beaumont, but few expressed any interest in St. Martin's welfare. Beaumont did not coerce St. Martin and indeed treated him well. Contemporaries found St. Martin's case to be far more defensible ethically than experimentation on animals, and no one mentioned St. Martin's rights in the matter. Some recent authors have even credited Beaumont with developing ethical principles of human experimentation, which include voluntary consent and lack of distress to the experimental subject.

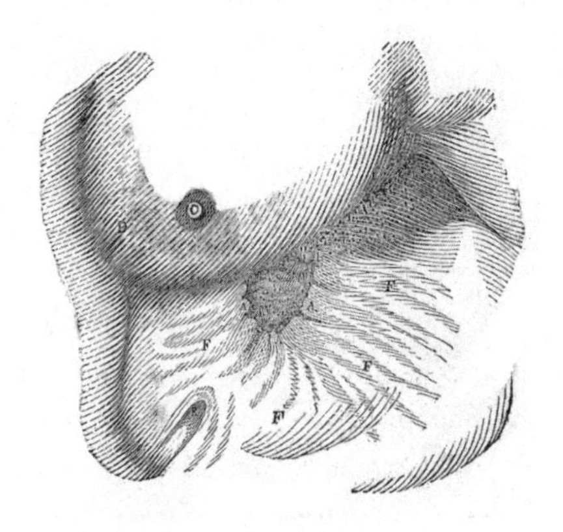

St. Martin's fistula, from William Beaumont, *Experiments and Observations on the Gastric Juice, and the Physiology of Digestion*, 1833. Wellcome Collection

But St. Martin's relationship with Beaumont was more ambiguous than these ideals imply. St. Martin's financial need and lower social status put him in a dependent relationship with Beaumont that was only partially voluntary, and Beaumont clearly put the interests of science before the interests of his experimental subject. To Beaumont and his peers, the master-servant relationship between the doctor and the fur trapper was a natural consequence of their differing social stations.

the entire animal kingdom. The aim of experimental physiology, he said, was "to conquer living nature, act upon vital phenomena and regulate and modify them."[8]

Claude Bernard admired William Beaumont's work with Alexis St. Martin, and in his *Introduction* he justified human experimentation in certain circumstances:

> It is our duty and our right to perform an experiment on man whenever it can save his life, cure him or gain him some personal benefit. The principle of medical and surgical morality, therefore, consists in never performing on man an experiment which might be harmful to him to any extent, even though the result might be highly advantageous to science. . . . For we must not deceive ourselves, morals do not forbid making experiments on one's neighbor or on one's self; in everyday life men do nothing but experiment on one another. Christian morals forbid only one thing, doing ill to one's neighbor. So, among the experiments that may be tried on man, those that can only harm are forbidden, those that are innocent are permissible, and those that may do good are obligatory.[9]

He left judgment of what was harmful or "innocent" to the experimenter.

Bernard advocated experimental physiology from intellectual conviction and personal circumstance. Despite being Magendie's right-hand man, he resembled Marshall Hall in having no official position or support for part of his career. Even after he was appointed to Magendie's chair, his laboratory space and resources were far less than he wished, especially in comparison with those of German scientists, whose large, well-equipped labs could support many students and researchers. Bernard's marriage to Marie-Françoise Martin in 1845 was one of convenience, for her wealth helped to support his research. But the marriage was unhappy, and Bernard and his wife soon became estranged. Later in life, she and their daughters became ardent antivivisectionists, much to Bernard's dismay.

Bernard's *Introduction to the Study of Experimental Medicine* was a manifesto for a new science as well as a plea for additional support from the French government. It also served as a wake-up call for French science. In the 1830s the Paris clinical school had been the model for modern science, but by the 1860s the German universities had usurped that position. Highly competitive German states poured money into their universities, creating laboratories and research institutes that were the envy of the world. Central to the German model was the role of the laboratory in training the next generation of scientists. When Germany unified into a single state in 1870—and shortly thereafter handed France a humiliating defeat in the Franco-Prussian War—it seemed an unbeatable power, and British and American scientists as well as French ones urged their governments to help them catch up. In the United States, the German model inspired the organization of the medical school of the Johns Hopkins University in Baltimore, founded in 1891, which in turn became the model for the reorganization of American medical education after 1900.

Although Bernard admired German success, his approach in the *Introduction* differed significantly from that of the more mechanistic Germans, who assumed that physiology could be reduced to chemistry and physics. Bernard insisted on the uniqueness of life and the uniqueness of physiology among the sciences. Vital phenomena followed laws—otherwise experimenting would be useless—but chemistry and physics could not entirely explain them. His model of experimentation began with observation and moved on to the creation of a hypothesis and experiments to test the hypothesis. He constantly returned to experimentation as he revised his hypotheses. Bernard viewed this as a highly rational process, but intuition also played a role in the development of a hypothesis.

He devoted much of the *Introduction* to vivisection: methods, choice of animals, and justification. "We shall succeed in learning the laws and properties of living matter only by displacing living organs in order to get into their inner environment," he wrote. "To learn how man and animals live, we cannot avoid seeing great numbers of them die, because the mechanisms of life can be unveiled and proved only by knowledge of the mechanisms of death."[10] Unlike Hall, who argued that "lower" animals were preferable, Bernard frequently experimented on dogs and insisted that the more closely the animal resembled humans, the more useful it was, although he refused to experiment on primates (see chapter 6). While dogs remained the most common experimental animal, researchers began to work with rats in the 1850s, and by the end of the nineteenth century rats and mice had become prominent, particularly in disease research.

While Bernard was aware of anesthesia and its uses—he wrote a treatise on anesthetics, based on extensive investigation on animals—he barely mentioned its use in the *Introduction.* He employed anesthesia at times, but more to calm an animal than to relieve its pain; curare, for example, would immobilize an animal but not relieve its pain. Manifestation of pain was part of the experimental process and did not call for special recognition or relief. Bernard saw himself as the successor to Galen and Harvey (his textbook on physiology has been called an updated version of Galen's *On Anatomical Procedures*), and he shared Galen's contempt for rivals and critics. The scientist was a special being, immune to the "cries of people of fashion" as well as to those of the animal; he saw only his scientific goal, and only his fellow scientists should judge him. "We," as scientists and as humans, had the right to vivisect animals, Bernard wrote, "wholly and absolutely."[11] It is not surprising that he succeeded Magendie as a target of antivivisectionists.

Frances Power Cobbe Fights for the Antivivisection Cause

By the middle of the nineteenth century a gap had opened between science and the nonscientific public. Bernard firmly closed the doors of his laboratory, and the public demonstrations of Magendie were a thing of the past. Also excluding the public was the rising specialization of science, with a proliferation of journals written in increasingly technical language. Yet in his *Introduction* Bernard deliberately addressed the public and challenged the antivivisectionists. While the French Société protectrice des animaux (Animal Protection Society), founded in 1846, had formed a committee to investigate vivisection in 1860, British protest against vivisection had been far stronger.

Lacking local targets, animal protectionists in Britain continued to look overseas. In 1846, a clergyman made the first of many protests to the French government over the treatment of horses at the veterinary school at Alfort. This issue came to the attention of the RSPCA in 1857, the same year in which Queen Victoria gave birth to her youngest child under chloroform anesthesia. Horses at Alfort were used for surgical training, not research, and were subjected to repeated operations without anesthesia. Both the popular and the medical press in Britain in the late 1850s and early 1860s uniformly condemned the practice, and the RSPCA organized a delegation to Emperor Napoleon III in 1861 that demanded an end to vivisection at the veterinary schools. A French commission that included Bernard recommended in 1863 that vivisection continue, but under more controlled conditions, including the use of anesthesia. The French Academy of Medicine, deeply resenting English interference, indignantly rejected the commission's findings.

The RSPCA's campaign against the French brought the cause of anti-

4.4 The True Vivisector

Thousands of copies of Frances Power Cobbe's antivivisection pamphlet *Light in Dark Places* (1888) were distributed for free. The illustration here, from Claude Bernard's textbook of physiology (*Leçons de physiologie opératoire*, 1879), appeared in Cobbe's pamphlet, accompanied by text from the Russian physiologist Elie de Cyon (1843–1912):

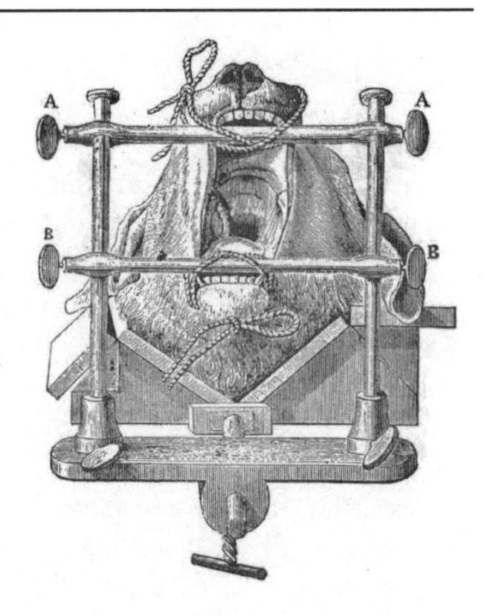

Illustration from Claude Bernard's *Leçons de physiologie opératoire*, 1879. Frances Power Cobbe, *Illustrations of Vivisection . . . from the Works of Physiologists* (Philadelphia, 1888), 9

The true vivisector must approach a difficult vivisection with the same joyful excitement, and the same delight, wherewith a surgeon undertakes a difficult operation, from which he expects extraordinary consequences. He who shrinks from cutting into a living animal, he who approaches a vivisection as a disagreeable necessity, may very likely be able to repeat one or two vivisections, but will never become an artist in vivisection. . . . The pleasure of triumphing over difficulties hitherto held insuperable is always one of the highest delights of the vivisector. And the sensation of the physiologist, when from a gruesome wound, full of blood and mangled tissue, he draws forth some delicate nerve-branch, and calls back to life function which was already extinguished—this sensation has much in common with that which inspires a sculptor, when he shapes forth fair living forms from a shapeless mass of marble.

vivisection to the attention of Frances Power Cobbe, who would become the driving force of the British antivivisection movement. Cobbe (1822–1904) embodied many of the characteristics and contradictions of the nineteenth-century movement. An upper-class Englishwoman, she worked as a journalist, writing on religious, feminist, and philanthropic topics. Forceful and energetic, she cultivated a wide range of friends and supporters from the clergy, the aristocracy, and the literary world and produced a flood of influential letters and articles.

Cobbe, in Florence in 1863, led a protest by the English community there against the work of Moritz Schiff (1823–1896), a professor at the Specola

Natural History Museum. From hearsay evidence, Cobbe concluded that Schiff not only experimented on animals but did so with unusual cruelty. She wrote up a petition, obtained almost 800 signatures, and sent off a letter of protest to the *Daily News* of London. She sent the petition to Schiff, who responded angrily in local newspapers, denying that any animals suffered unduly, assuring critics that he and his associates employed anesthesia, and arguing for the scientific necessity of vivisection. British newspapers took far more interest in the issue than did Italian ones, and the case, like earlier protests against French practices, displayed the conviction that cruelty to animals, especially dogs, cats, and horses, was simply un-English.

In this era of anesthesia and heightened consciousness of pain, experiments that deliberately caused pain were an obvious focus for antivivisectionists. But pain was not the only issue. Evolutionary theory, recently enunciated by Charles Darwin (1809–1882) in *On the Origin of Species* (1859), indicated that humans and animals shared common ancestors. Perhaps more important for nineteenth-century urbanites, animals represented a connection with nature and a rural past that was rapidly disappearing, while science foreshadowed a modern world of ceaseless change and instability. Older values appealed to Cobbe's allies, the clergy and the aristocracy, yet as an independent, self-supporting, and unmarried woman, Cobbe also represented new values.

Cruelty may have been un-English, but by 1870 British physiologists had begun to adopt Continental models of experimentation, including vivisection. The publication of John Burdon Sanderson's *Handbook for the Physiological Laboratory* in 1873 marked a turning point. Sanderson (1829–1905), a professor of physiology at the University of London who had studied with Bernard in Paris, offered the first detailed laboratory manual in English of the methods of French and German physiologists. Not only did the *Handbook* have an immediate impact on British physiology; it also had an immense impact on the British antivivisection movement. Here in print, with copious illustrations, was an animal lover's worst nightmare.

The combination of Sanderson's book, a renewed attack on Moritz Schiff, and a vivisection demonstration at the annual meeting of the British Medical Association in 1874, once more brought the issue of experiments on animals before the British public. Articles and editorials poured forth. Cobbe concluded that new legislation was the only way to restrict animal experimentation (commonly referred to as vivisection, although not all of it

Caricature of Charles Darwin as an ape, published in the satirical magazine *The Hornet*, 1871. Wikimedia

4.5 Darwinian Animals

Soon after he returned in 1836 from his five-year voyage on the *Beagle*, Charles Darwin wrote in his notebook, "Animals our fellow brethren in pain, disease, death & suffering, & famine,—our slaves in the most laborious works, our companions in our amusement—they may partake of our origin in one common ancestor; we may all be netted together."

In *The Origin of Species* (1859), Darwin proposed his theory of evolution by natural selection, making it clear that humans had evolved from other related species, particularly apes. In *The Descent of Man* (1871) he wrote, "The main conclusion arrived at in this work, namely that man is descended from some lowly organized form, will, I regret to think, be highly distasteful to many." Ultimately, however, Darwin's de-centering of humans became a powerful philosophical argument for the essential equality of humans and animals.

■ Quotes from Charles Darwin, *Notebook B: [Transmutation of species (1837–1838)]*, CUL-DAR121, 231–32, transcribed by Kees Rookmaaker (*Darwin Online*, http://darwin-online.org.uk/), and Darwin, *The Descent of Man*, 2 vols. (London: John Murray, 1871), 2:404.

was) in Britain. Richard Martin's Act of 1822, even as subsequently amended, was not strong enough. Cobbe urged the RSPCA to act, but its members could not agree on how to proceed. Finally, she took matters into her own hands, and her agitation in summer 1875 led to the appointment of the Royal Commission on the Practice of Subjecting Live Animals to Experiments for Scientific Purposes.

The RSPCA remained divided, and in 1875, while the royal commission deliberated, Cobbe founded the Victoria Street Society for the Protection of Animals from Vivisection. The Victoria Street Society, whose logo was a faithful, sad-eyed dog, marked the beginning of the modern antivivisection movement. Cobbe used her many connections to gain patronage and prestige for the society, and the poets Alfred Tennyson and Robert Browning were among its early supporters. The society sponsored public meetings and petitions to lobby the royal commission. The commission recommended regulatory legislation, but Cobbe's allies introduced a far more restrictive bill in the spring of 1876. This bill allowed vivisection only when the knowledge to be gained was of clear medical benefit to humans, and dogs and cats were entirely exempted. Scientists, newly self-conscious following the commission, lobbied heavily to modify this bill, and even though Queen Victoria herself barraged government ministers with antivivisection memos, the bill was modified in favor of the scientists. Nonetheless, the Cruelty to Animals Act of 1876 was the first attempt by a national government to regulate animal experimentation. It was not replaced by new legislation until 1986. The 1876 act required the Home Secretary to license and register all experimenters and to register and inspect all places of experiment. Experiments for teaching, as well as experiments without anesthesia on dogs, cats, horses, mules, and donkeys, required special certification. Applications for licenses and certification required endorsement by a recognized scientific authority, such as the Royal College of Physicians of London. Fines were established, and inspectors appointed.

Cobbe and other antivivisectionists viewed the 1876 act as a sellout to the scientists. From this point on, scientists and antivivisectionists would become increasingly polarized, with animal protectionists such as the RSPCA occupying an uneasy middle ground. In 1876 the antivivisection movement had barely begun, and following the act it quickly became radicalized. In 1878 the Victoria Street Society declared that it would fight for the total abolition of vivisection, severing its ties with more moderate reformers. Cobbe continued her attempts to introduce new bills into Parliament, with little success. British physiologists continued to reorganize their enterprise on the Continental model. In retrospect, 1876 was the last opportunity for compromise between scientists and antivivisectionists. In 1898 the Victoria Street Society ousted Cobbe as its president and changed its policy from the abolition of vivisection to more moderate goals of restriction and regulation. Cobbe founded the Union for the Abolition of Vivisection, but the movement had relegated her to its fringes. Meanwhile, the new science of bacteriology, which relied heavily on animal experimentation, began to give human and veterinary medicine the power to heal (see chapter 5).

The Special Feelings of Women

Although men usually occupied the highest positions in the antivivisection movement, at least half of antivivisection activists were women. In a period when women were largely barred from public life, this was remarkable. Why were so many women involved? The notion of "separate spheres" had relegated middle- and upper-class women to the household, differentiating them sharply from the industrial working class. Frances Power Cobbe urged these women to work against vivisection with both traditional and feminist ideology. Women, she said, were naturally more sensitive and spiritual, and therefore more attuned to the plight of animals. Yet women's higher moral character made it essential that they enter public life and not just operate from the home. Cobbe persuaded women who were put off by feminism and the suffrage movement (in which Cobbe was also involved), showing them that they need not sacrifice their "womanliness" to be active, effectively countering arguments used by opponents of female suffrage. Cobbe's arguments also persuaded women elsewhere in Europe and in the United States to join the antivivisectionists.

For feminists such as Cobbe, antivivisection offered another outlet for their energies, giving them valuable experience. Other women had differing motives. Some, seeking charitable causes, found antivivisection attractive. For others, antivivisection held connections with earlier movements for the abolition of slavery: Caroline Earle White (1833–1916), founder of the American Anti-Vivisection Society in 1883, was a daughter of the abolitionist lawyer Thomas Earle. In France, Claude Bernard's estranged wife and their daughters rescued experimental animals, especially dogs, and cared for stray dogs from the streets of Paris. A probably fictional story tells of a search by Bernard's daughter for her friend's pet dog, only to find it in her father's laboratory. Mme Bernard and her daughters spent much of their fortune on animal causes, and Mme Bernard was a longtime member of the French antivivisection society.

The physicians Elizabeth Blackwell (1821–1910), the first woman to graduate from an American medical school, and Anna Kingsford (1846–1888) saw parallels between the rise of surgery on women, especially gynecological surgery, and vivisection. While in medical school, Blackwell and Kingsford had witnessed public hospitals using poor women as teaching tools for student doctors. At a time when showing the ankles was considered immodest, these two women were horrified to see other women's bodies exposed to the cynical gaze of male medical students. Because, as we have seen, the poor and nonwhites were thought to be less sensitive to pain, surgical operations, even in the 1880s, were performed without anesthesia. Anna Kingsford explicitly compared the poor, women, and animals—all considered fit subjects for painful experiments. Many physicians, both male and female, had doubts about the rise in such surgeries as removal of the ovaries or the clitoris to solve "women's problems," which could include "hysteria." Kingsford and Blackwell connected this to vivisection, and Blackwell referred to the removal of ovaries as "castration." Blackwell saw only the thinnest of lines between animal vivisection and experimentation on poor women. Kingsford was more extreme, offering herself as a replacement for experimental animals and cursing Claude Bernard and other experimenters. The patrician Cobbe felt that Kingsford's behavior only reinforced the notion of the overly emotional woman. The association of antivivisection with women endured: in the 1920s, Morris Fishbein, editor of the *Journal of the American Medical Association*, asserted, "Isn't it after all a question of sparing the hyperesthetic sensibilities of some idle woman rather than the duller sensibilities of some lower animal?"[12]

Fictional works such as the popular *Black Beauty* (1877), a biography of a horse written by the English Quaker Anna Sewell (1820–1878), and *A Dog of Flanders* (1872), by Ouida (Maria Louise Ramé, 1839–1908), dramatized animal suffering. In the 1890s, two bestselling works of fiction portrayed the scientist in widely differing ways, betraying the deeply conflicted feelings of the public over scientific advances. Arthur Conan Doyle (1859–1930), a physician, introduced Sherlock Holmes in 1887, white-coated, in a laboratory; the short stories that made him famous appeared in *The Strand Mag-*

azine from 1891 to 1893. Holmes did not hesitate to prick himself in the finger to obtain a blood sample, and he regarded humanity with a cold and clinical gaze. H. G. Wells (1866–1946), who had studied science at the University of London and authored a biology textbook, presented the antivivisectionist's nightmare in *The Island of Doctor Moreau* (1896). Although his fellow scientists had ostracized Moreau for his horrific vivisections, the implication remained that science, unrestrained, was capable of still greater horror unless nature herself rebelled.

In his 1875 defense of animal experimentation, the American physiologist John Call Dalton (1825–1889) could point to increased knowledge of human physiology but few practical medical advances. The antivivisection movement based its arguments on the notion of needless pain, for scientists had not demonstrated to the public that their experiments were worthwhile. As we will see in the next chapter, twenty years later this picture had changed greatly, and the advent of microbiology promised to eradicate human disease for good, with the help of the bodies of millions of animals.

5 The Microbe Hunters

In 1878, the year Claude Bernard died, a Frenchman coined the word *microbe* to denote a microscopic infectious agent. *Microbe* soon encompassed a great variety of organisms, including bacteria, protozoa, and fungi, which caused infection in different ways. Other infectious agents for diseases such as smallpox and rabies could not be found and were presumed to be too small to be viewed under conventional optical microscopes. These agents were viruses, which are much smaller than other disease-causing entities. The adoption by scientists and physicians of the germ theory of disease, which stated that contagious diseases were caused by the action of microbes, opened new arenas for research that promised immediate human benefit. Although Bernard and other physiologists promised that their research on animals would lead to the improvement of human health, they placed these goals far in the future. While physiology and microbiology differed, each discipline depended on human and especially animal bodies for experimental proof.

By 1898, when the Victoria Street Society ousted its president, Frances Cobbe, the promise of science to improve human health had become a reality. New microbes were being discovered almost daily, and a new vocabulary of vaccines, immunizations, and antitoxins entered everyday life. Antisepsis and asepsis, together with anesthesia, revolutionized surgery, making complex operations possible and elevating surgeons to the elite of the medical profession. Dozens of scientists contributed to these developments, but two men were largely responsible for the development of the germ theory of disease in the 1870s and its subsequent public acceptance: the French chemist Louis Pasteur (1822–1895) and the German physician Robert Koch (1843–1910). Employing thousands of animals and a few human subjects, they developed the theories and techniques that received their fullest expression in the years before World War I in the Frankfurt laboratory of Paul Ehrlich (1854–1915). With Ehrlich's work, the laboratory decisively joined the clinic in the task of preventing and treating disease, and "scientific medicine" became a reality. Yet, as is always true of science, this process was neither straightforward nor inevitable, and the germ theory had many critics. Despite the heroic stature subsequently granted the founders of microbiology by press and public, ambiguities and ethical lapses were also part of the process.

Pasteur and Microbes

More than a century after his death, Pasteur is undoubtedly still the most famous scientist in France. Every French village has a rue Pasteur. Yet Pasteur's early career made him an unlikely candidate for the founder of bacteriology. The son of a provincial tanner, Pasteur trained as a chemist, not a physician. His early studies of the optical qualities of the crystals of certain organic acids led to the interesting discovery that some crystals were not symmetrical, having left- and right-handed versions. Pasteur found that only the molecules of living substances exhibited this asymmetry. Thus, he came to the study of biological problems from the study of chemical problems; but he came convinced, like Bernard, that biology somehow differed from the physical sciences.

In the 1850s Pasteur held an appointment at the University of Lille, in northern France, where the distillation of juice from sugar beets into alcohol was an important local industry. Pasteur began to study the process of fermentation, by which alcohol was produced from the juice. Fermentation helped to turn grapes into wine, flour into bread, milk into cheese, but no one knew how it worked. Fermentation was related to putrefaction: rotting was the next step in the process, and it often occurred despite efforts to prevent it. By means of chemical analysis and microscopic investigation of fermenting beet juice, Pasteur concluded that fermentation was not an inorganic chemical process. Its cause was a living organism, microscopic in size, called *yeast*. Pasteur's fellow chemists were not happy with this mingling of inexact biology with rational, quantitative chemistry. While physiologists were endeavoring to take the messy uncertainty out of biology by reducing it as much as possible to physics and chemistry, the chemist Pasteur put uncertainty back in. In 1859 the Paris Academy of Sciences awarded Pasteur its prize in experimental physiology for his work on fermentation. The chair of the prize committee was Claude Bernard.

The implications of Pasteur's explanation of fermentation for the theory of disease were not immediately evident, least of all to Pasteur. In Paris in the 1860s, he took on the problem of spontaneous generation, the age-old question whether life could come from nonlife. Contemporary theorists argued that a vital principle in nature could spontaneously organize inorganic substances into living ones. Pasteur argued that only living microorganisms, many of which were in the air, could produce more organisms. Through a series of experiments involving infusions in sealed and unsealed flasks, Pasteur demonstrated that the flasks left open to the air generated microorganisms, while those heated and sealed from the air did not. In fact, there was a certain amount of luck in Pasteur's success, since as he later found, some microorganisms could withstand heat and even seemingly clean flasks could be contaminated.

In the mid-1860s, Pasteur was asked to investigate an epidemic of a disease known as *pébrine*, which attacked silkworms (the caterpillars of moths). Silk was a major industry in France, and rescuing the silkworm was of national importance. However, Pasteur did not yet understand how microorganisms could cause disease. By the example of fermentation, a microorganism might cause chemical changes in its host but not kill it. Contemporary medical debates on disease causation centered on the idea of a *miasm* or atmospheric condition that could bring on certain diseases. Many countries implemented public health measures, including cleaning streets, purifying the water supply, and providing sanitary waste disposal, in order to stem the tide of epidemic diseases such as cholera and tuberculosis, which flourished in the damp and dirty conditions of industrial cities. Microorganisms played little role in this picture.

Pasteur soon found microorganisms in silkworms. At first he believed the "corpuscles," as he called them, were only the symptoms of pébrine, and that they only came into view under the microscope when the disease was advanced. He came to recognize that the microorganism caused the illness and that it did not travel through the air but was transmitted through the moths' eggs. At the same time, he identified another silkworm disease known as *flacherie* and concluded that it too was caused by a microorganism, in this case one that flourished on damp mulberry leaves, the preferred food of silkworms. Hygienic measures, such as keeping the silkworms' beds clean and dry, prevented the disease. In both cases, he showed that a specific microorganism caused a specific disease.

Next, he developed the process of *pasteurization*, which killed certain microorganisms that caused spoilage by applying heat for a specified period. This process was first used for wine—another major French industry—and then for beer and, later, milk. In the 1860s, the British surgeon Joseph Lister applied Pasteur's ideas about microbes to the phenomenon of post-surgical infection and introduced antiseptics to kill microbes on tools and in the air, which made surgery much less risky. Aseptic practice—preventing infection through sterile techniques and tools rather than with chemical antiseptics—was even more effective. Pasteur himself promoted sterile techniques and experimented on dogs with different kinds of wound dressings.

In 1865, Pasteur and Bernard were members of a commission to study cholera, which was about to reach epidemic proportions in Paris. Convinced that a microorganism caused the disease, Pasteur had no idea how it acted, and he believed that it was carried through the air (it is actually carried by food and water). He analyzed the air of a cholera ward, as well as dust from the floor and bedding, but failed to find an infectious agent. Bernard took blood samples, which he analyzed chemically, rather than seeking microbes. A dozen years later, when Pasteur began to investigate the animal disease anthrax, he recognized that the air might not be the only vehicle of infection.

In 1850 two Frenchmen, the physician Casimir-Joseph Davaine (1812–1882) and the pathologist Pierre François Olive Rayer (1793–1867), had noticed a rod-shaped microbe in the blood of anthrax victims. A decade later, following Pasteur's fermentation experiments, Davaine injected sheep's blood that he believed to be infected with anthrax microbes into rabbits, which soon died. Others pointed out that the rabbits might not have died from anthrax; cow's blood with no trace of the microbe was equally fatal. In 1877, as Pasteur began to work on anthrax, a rural German physician named Robert Koch grew the anthrax bacteria outside the body, using the aqueous humor of an ox's eye as the medium. Under the microscope, he could see the characteristic rods grow and multiply.

Koch was fortunate. Anthrax is the largest disease-causing bacterium and the easiest to see and distinguish from other microorganisms. In addition, Koch's new staining techniques made the anthrax rods even more evident. Anthrax attacked economically valuable domesticated animals such as cattle, sheep, and goats, and it was a *zoonosis*, an animal disease that easily crossed the species barrier to afflict humans as well, particularly humans who had contact with farm animals. An epidemic of anthrax had killed half the sheep in Europe in the early eighteenth century, and those who handled animal skins and fleece were especially susceptible to contracting the disease by inhaling its spores; it was sometimes known as "woolsorters' disease." It was therefore an ideal disease for the first demonstrations of the germ theory.

In 1877, Pasteur and Koch each published on the *etiology*, that is, the causes or origins, of anthrax, describing its life cycle, how it got into the blood, and how it caused disease. This was the first widely accepted demonstration of a bacterial cause of disease. In a paper on wound infections published in the next year, Koch discussed the procedures he had followed, which came to be known as "Koch's postulates":

1. The microorganism in question should be found in each case of the disease, that is, in each experimental animal.
2. The microorganism should not be found in cases of other diseases (specificity).
3. The specific organism should be isolated.
4. The organism should be *cultured*, that is, grown outside the body in a sterile medium.
5. When inoculated into a well animal, the organism should produce the same disease.
6. The same specific organism should then be recovered from the diseased animal.

These postulates emphasized the cultivation of the suspected microbe outside the human or animal body in the controlled environment of a sterile artificial medium. Koch himself used a variety of gelatin. He also used animals to retrieve and to test the specificity of microorganisms, reasoning that they were receptacles for the microorganism just as the dishes of gelatin were. In developing vaccines for anthrax, chicken cholera, and swine fever from 1878 to 1883, Pasteur followed Koch's techniques, with great success. In each of these cases, Pasteur developed a vaccine by weakening the virulence of the microbe, whether by heat, chemicals, or prolonged exposure to the air. The principle of a vaccine, he believed, must follow the model of Edward Jenner's smallpox vaccine: a weaker variety of the disease in question could provoke immunity.

He cultivated the microbe in a sterile medium. But in these studies,

Pasteur and his assistant, the physician Émile Roux (1853–1933), discovered that the virulence of a microbe could also be altered—either increased or decreased—by passing it through living animal bodies. Pasteur passed attenuated (weakened) cultures of chicken cholera through a series of chickens to increase its virulence. Koch had also noticed this phenomenon and thought that the repeated passages acted to purify the microbe. Pasteur gradually came to realize that the microbe itself changed—what is now called *mutation*.

The Rabies Vaccine

Rabies is a nasty animal disease caused by a virus that humans occasionally contract, generally by being bitten by a rabid animal. It attacks the central nervous system and sometimes causes an inability to swallow—thus its alternative name, *hydrophobia*, "fear of water." Although in the nineteenth century rabies was uncommon both in animals and in humans, it excited particular dread for several reasons: its long incubation period (several months in humans) meant an extended period of anxiety after possible exposure; its symptoms were horrific, painful, and invariably fatal; and a common carrier was the domestic dog, which in this period was consolidating its status as the companion animal of choice.

In economic terms, an anthrax vaccine was more valuable than one for rabies. Vastly more economically valuable animals such as cattle and sheep died of anthrax. Very few people died each year of rabies, and even if one were bitten by a mad dog, rabies did not invariably result. But the high public profile of rabies meant that Pasteur gained enormous fame for his vaccine. The dramatic story of its first human trial, on the boy Joseph Meister, added to Pasteur's public acclaim.

Pasteur and Roux began their quest for a rabies vaccine in 1880. Pasteur could not find the rabies microbe since, as a virus, it remained invisible with his optical microscope, but he continued to be convinced that the cause of rabies was a microbe and that the techniques he had used to develop vaccines for other animal diseases would also be effective with rabies. Pasteur at first believed that immunity could only be conveyed by a living but weakened microbe, which would consume the nutritional supplies of the host organism and leave nothing for the more virulent microbes that followed. Later he concluded that the microbe released a chemical toxin and that immunization was therefore a chemical process. If this seems like a surprising slip for Pasteur the chemist, it also serves to remind us of the immense uncertainty surrounding microbes and their actions in this period, as well as the vast differences among microbes. Pasteur thought he understood what caused immunity, but he did not know how the body produced immunity. While today the phrase *germs cause diseases* seems too obvious to need explanation, 140 years ago every word in that phrase was questionable: What was a germ? What was disease? How was disease caused?

Building on their previous experience, Pasteur and Roux employed animal bodies rather than sterile gelatin in their quest for a rabies vaccine. They had already noted that passing a disease through several individuals of the same species could markedly affect its virulence. After a certain number of passages, the microbe would become "fixed," or stable. In addition, a French veterinarian reported in 1879 that when rabies was transmitted from dogs to rabbits, the month-long incubation period for the disease seen in dogs dropped to about eighteen days in rabbits. Pasteur found that additional passages through rabbits reduced the period even further, to six or eight days. In addition, rabbits and guinea pigs, which he also used, were more docile than rabid dogs.

Pasteur's decision to use animals as his test tubes had momentous consequences for the future use of animals in research. Thereafter, bacteriological and immunological research became inextricably linked to the use of animals as culture media. As the scale of research grew, so too did the numbers of animals used. Pasteur used hundreds of animals to develop the rabies vaccine; from 1882 to 1885, for example, he passed the virus through 90 sets of rabbits. Paul Ehrlich, 20 years later, used thousands of mice to develop Salvarsan, as we will see below. The development of the polio vaccine in the 1950s, discussed in chapter 6, required the sacrifice of millions of monkeys. Physiological research, by contrast, employed a fraction of these numbers.

Pasteur began his rabies research as he had his other disease research: he injected animals with different substances—saliva, blood, or tissue from rabid dogs—and then waited to see what happened. He discovered that nervous tissue, and especially the brain, was the ideal medium for cultivating the rabies microbe, and he claimed that injecting rabid brain matter into animal brains invariably caused the disease. He also used monkeys, which at the time were rarely used for research, finding that while passage of rabies through rabbits made the virus more virulent to dogs and presumably humans, passage through monkeys made it less virulent. A sufficient number of passages through monkeys would weaken the microbe enough, he surmised, to make an effective vaccine, and in the summer of 1884 he proclaimed that he had tested such a vaccine on dogs and found it effective.

Pasteur worked on several other methods of attenuation. One method involved drying the spinal cords of affected rabbits. To immunize the dogs, he injected them with portions of spinal cords of gradually increasing virulence. It should be noted, however, that the nature of immunity and its acquisition was so imperfectly understood that at first he injected them with cords of gradually *decreasing* virulence.

In 1885 Pasteur was on the verge of success. He continued his experiments in inducing immunity in dogs and began experiments to test whether the method of immunization would work even after the victim had been exposed to rabies, with its long incubation period. If it were possible to inoculate a victim soon after exposure and prevent the disease from manifesting itself, then Pasteur's vaccine could also act as a cure for a dreaded disease. According to the historian Gerald Geison, who closely examined Pasteur's laboratory notebooks, the scientist attempted to cure two humans by his method of inoculation in the spring of 1885. In both cases the required series of injections were not completed, and the results were inconclusive. Up to that point, experiments on exposed dogs were also inconclusive.

On a hot July day in 1885, 9-year-old Joseph Meister, the son of a baker in eastern France, sat trembling in the office of the famous Pasteur. He was weary from the long train trip to Paris, and the many dog bites on his hands, arms, and legs hurt horribly. The local physician had cauterized some of the bites with phenic acid, a standard treatment, which had only made them hurt more. Monsieur Vone, the grocer, who owned the dog, had come with them on the train. The dog had also bitten him but not as severely. He had killed the dog, who was obviously mad. Everyone in the village said that Pasteur was Joseph's only hope, that without Pasteur's help he would die a terrible death.

Pasteur brought two physicians in to see Joseph, and they agreed that given the number and severity of the bites, the boy would almost certainly develop rabies. Pasteur made a decision. As he wrote in his paper of October 1885 describing the case, "The death of this child appearing to be inevitable, I decided, not without lively and sore anxiety, as may well be believed, to try upon Joseph Meister the method which I had found constantly

Pasteur with the first Russian children treated with his rabies vaccine, ca. 1889. Bibliothèque interuniversitaire de santé, Paris, Licence ouverte

5.1 The Rabies Virus

The Pasteur Institute became a destination for victims of dog bites from across Europe and beyond.

successful with dogs."[1] But according to the evidence of his notebooks, Pasteur had not in fact completed the relevant experiments on dogs. Nonetheless, he decided to go ahead and treat Joseph. Over a period of ten days, he injected the child with increasingly virulent preparations of dried rabbit spinal cord, ending on 16 July with the most virulent. Joseph did not come down with rabies—either with the natural version he had acquired or with the injected rabbit version, which would have manifested itself much more quickly. In mid-August Joseph was declared cured, and Pasteur became a national hero.

As his paper was being published in October, Pasteur was treating another young rabies victim, a shepherd named Jean-Baptiste Jupille. He also did not contract rabies, and by the end of 1885 a portion of Pasteur's laboratory had become a clinic for dog-bite victims. In 1889, the Pasteur Institute opened, the first publicly endowed research laboratory in France. Both Meister and Jupille later became guards at the institute.

Was Pasteur's treatment of these boys unethical? By modern standards, absolutely. To perform human trials without adequate demonstrations on animals is a violation of modern clinical and experimental ethics; had Pasteur been in a modern laboratory, answerable to institutional ethics committees, he would not have been allowed to go through with his treatment. Even his assistant, Émile Roux, believed that Pasteur had inadequate animal evidence on which to base his conclusions, and he refused to inject Meister or the others (since Pasteur was not a physician, he could not perform the injections himself). On the other hand, Joseph Meister seemed to face certain death. We now know that not everyone exposed to rabies develops the disease, but at the time opinion was unanimous that the boy would die without treatment. In addition, neither Meister nor his mother nor Jupille could be said to have been adequately informed, by modern standards, of the possible results of their treatment. The ethical dilemma of Joseph Meister's case presents on several levels. Did Pasteur believe he was acting ethically, or was he motivated by ambition? Are these motives mutually exclusive? Was Roux correct to assert that no human trials were possible given the current state of knowledge? Can we apply modern ethical standards to activities in the past? With present restrictions on research, could Pasteur have made his discoveries today? These questions are not easily resolved.

Koch and the Antivaccinators

At the time of his 1877 paper on anthrax, Robert Koch was an obscure German physician in what is now Poland. Working alone in a back room of his house, Koch developed experimental techniques critical to the development of bacteriology. He used the microscope extensively, but one of the problems with looking at bacteria under the microscope, especially ones smaller than the relatively large anthrax rods, was their lack of contrast against the tissue background. Koch developed methods of staining, fixing tissue specimens, preparing slides, and especially photographing specimens. His published papers were among the first to employ photographs. To skeptics of the germ theory, these photographs were especially convincing, showing that different microorganisms looked different and supporting the theory that specific microbes caused specific diseases.

As we have seen, Koch also developed in vitro culture techniques. He grew the disease organism in an artificial environment of a sterile medium.

By this means, Koch could demonstrate how a microorganism reproduced and could then induce disease in a well animal—which could vary from a cow to a rat—by injecting the microorganism. Natural infections, especially wound infections, which he studied in the late 1870s, were far too variable to provide reliable proof. By isolating a single disease-causing set of microbes, Koch eliminated the chance infections that could destroy an experiment.

In 1880, the German government gave Koch a laboratory in Berlin. He left his medical practice for research. In contrast, Pasteur was not given his own research institute until 1889, at the end of his career. The German government's support of science meant that until World War I it led the world in microbiology. Koch inaugurated his laboratory by tackling tuberculosis, one of the most baffling and deadly diseases of the era. Tuberculosis was known primarily as a disease of the lungs, where its most evident symptoms resided, and nineteenth-century literature is filled with wasting, feverish tuberculosis victims, whose bloody coughs signaled a death sentence. Yet tuberculosis could also reside in a seemingly well human or animal for many years without provoking symptoms. There was no known cure for tuberculosis, and much debate about its causes. Its trail of contagion was difficult to trace, since one could contract the disease through contact with someone who displayed no symptoms.

Koch succeeded in isolating the tuberculosis bacillus by staining tissue from tuberculosis patients. His 1882 paper on tuberculosis created a sensation. Koch went through his postulates and described several experiments on animals. He demonstrated that like anthrax, the tuberculosis bacillus had a spore stage, during which it could be transported through the air and remain viable in a dry atmosphere, which normally would have killed it.

Although Koch showed that tuberculosis could be carried by milk, he did not believe that cows could transmit bovine tuberculosis to humans via their milk. Around 1900, the American bacteriologist Theobald Smith (1859–1934) demonstrated that children, especially, could indeed be infected with bovine tuberculosis by drinking contaminated cow's milk, and his results led to the widespread pasteurization of milk and more sanitary farming conditions, which helped check bovine tuberculosis. Although Smith trained in human medicine, he worked for many years at the Bureau of Animal Industry (BAI), established by the US Department of Agriculture in 1884 to research diseases of agricultural animals. His experience with animal diseases would prove critical in this and other cases.

However, attempts to develop a tuberculosis vaccine failed. Koch announced in 1890 that he had developed a substance called *tuberculin*, which cured tuberculosis. He did not reveal its composition but claimed its effectiveness in animal trials. However, tuberculin, a preparation of cultures of tuberculosis bacilli, proved to be ineffective and dangerous. The French physicians Léon Charles Albert Calmette (1863–1933) and Camille Guérin (1872–1961) developed a live bacillus tuberculosis vaccine in 1906 that is still employed in many countries, although it is little used in the United States. The BCG vaccine, as it is known, carries some risk of infection, and those vaccinated then test positive for tuberculosis, making detection of the natural disease difficult.

From 1880 to 1900, the microbial causes of several diseases were isolated, and the germ theory began to gain acceptance. Yet scientists disagreed about many issues, including how infection occurred, how the body resisted infection, what immunity was, and how vaccines worked. Pasteur's rabies vaccine gained much public attention, but so too did Koch's failure to develop a tuberculosis vaccine or cure. The discovery of the malaria parasite in 1880 meant that physicians could use the centuries-old medication quinine only for malaria, against which it was effective, and not for other fevers, and tests to detect the presence of other disease-causing microbes made diagnosis more accurate. Diseases seemed ever more variable. Laboratory techniques, including animal use, continued to be developed and refined, and microbiology firmly established itself as a discipline between the laboratory and the clinic.

A major success was the development of diphtheria and tetanus antitoxins in Koch's laboratory by Shibasaburo Kitasato (1852–1931) and Emil von Behring (1854–1917). Extending Émile Roux's work on diphtheria, they found that diphtheria and tetanus, two very different diseases, worked by essentially the same mechanism: the bacteria multiplied in a local area and produced a poison or toxin, which then circulated in the body—in diphtheria, via the bloodstream; in tetanus, via the nerves. They showed that they could induce tetanus in a well animal by injecting it with the bacteria-free (and cell-free) serum of a sick animal, usually a guinea pig: clearly the toxin alone caused the disease symptoms. Following Pasteur's principles of inducing immunity as well as principles first discovered by Mithridates, Kitasato and Behring then showed that animals injected with sublethal amounts of toxin became immune and could after a time survive a lethal dose. Moreover, other animals injected with the serum of immune animals gained immunity. But antitoxin was a therapy, not a vaccine. It could not control the spread of disease from person to person, and the injected antibodies did not last long in the body.

Many people remained unconvinced of the value of the germ theory. The miasma theory pointed to dirt, overcrowding, bad air, and unclean water as sources of disease, without reference to the action of microbes. Bad smells and dirt could themselves cause disease. Public health advocates noted that improved sanitation greatly reduced the incidence of diseases such as cholera. The inconsistent behavior of microbes and the failure to isolate microbial causes for all diseases (viruses continued to be poorly understood, and as we will see, deficiency diseases did not have microbial causes) discredited the germ theory in some eyes. In 1884 Koch isolated the bacillus that causes human cholera, but he could not then induce the disease in animals. Anticontagionists therefore argued that cholera was not contagious. In the 1892 cholera epidemic, scientists conclusively showed the presence of the cholera microbe in the water supplies of cholera-ridden cities, but some remained unconvinced. The German health reformer Max Pettenkofer (1818–1901) swallowed a glass full of microbe-laden water and did not fall ill; he proclaimed triumphantly that the germ theory was wrong. He was amazingly lucky. The Russian composer Peter Ilyich Tchaikovsky did the same thing, although not on purpose, and died of cholera.

Prominent opponents of the germ theory, particularly in Britain, were also strong antivivisectionists. A well-known physician, Sir Benjamin Ward Richardson (1828–1896), argued that a clean environment, clean bodies, and clean minds would eliminate disease, while experimentation on animals would not. Frances Power Cobbe would have agreed with these sentiments, although she also thought that undue attention to one's health was mere selfishness. The language of purity and impurity in the public health movement also condemned the introduction of foreign substances such as vaccines into the body. How could injecting diseased tissue—particularly diseased animal tissue—make you well?

The playwright George Bernard Shaw (1856–1950) offered a witty but scathing attack on the germ theory, vaccines, vivisection, and the medical profession in his 1913 play *The Doctor's Dilemma.* In a lengthy introduction, Shaw, a prominent Socialist and antivivisectionist, skillfully exposed popular fears and the pretensions of scientific medicine. The poor and the ill, he said, needed better food and housing more than medicine, and the new scientific medicine was designed not to keep people healthy but to cure diseases. Having lived through the vogue for public health, followed by the vogue for germs, Shaw was skeptical about new theories: "I have heard doctors affirm and deny almost every possible proposition as to disease and treatment."[2] How can doctors say that all diseases are caused by germs when they have not seen the germs of many of them? The germ of smallpox, after all, had still not been seen. Was not bacteriology mere belief and superstition, then? Common sense and cleanliness should win the day.

With the anticontagionists, Shaw believed that immunizations were dangerous and misguided. Koch's tuberculin fiasco showed that doctors followed the latest trend with little knowledge of the actual causes of disease. Shaw characterized immunizations as a plot to make money for physicians, claiming that most people could not afford them. He characterized vivisection as another fashion that physicians slavishly followed. The right to knowledge, he argued, did not include the right to cause pain, and indeed doctors above all should not cause pain. Benjamin Ward Richardson also made this argument and developed several new kinds of anesthesia to minimize animal and human pain.

Shaw did not spare upper-class antivivisectionists who criticized scientists but did not hesitate to hunt, eat meat, or wear furs. His conclusion that medicine should be socialized probably did not sit well with some, but Shaw expressed widely shared doubts and confusion about the new scientific medicine. Was medicine about to abandon the bedside for the laboratory, or could the two coexist? A century later this is still a relevant question.

Ehrlich and Salvarsan

Shaw's skepticism about scientific medicine was being challenged even as he wrote. In 1907, Paul Ehrlich patented a substance he called 606, later known as Salvarsan. Number 606, he reported, could cure syphilis. It was called 606 because it was the 606th compound Ehrlich had tried. Only in 1905 had two German scientists identified the corkscrew-shaped microorganism of the variety known as spirochetes, whose scientific name became *Treponema pallidum.* In the previous two decades several other microorganisms had been identified as the cause of syphilis, but the treponemes were found at all stages of syphilis and in all affected organs. In the next year, another German, August von Wassermann (1866–1925), gave his name to the Wassermann test, which accurately diagnosed syphilis from the presence of antibodies in the blood.

The origins of syphilis in Europe continue to be debated. Some argue that Columbus brought it back from the New World; others, that it was already in Europe. In any case, around 1495 a virulent, often fatal form of syphilis began to spread in Europe, and it remained the most serious sexually transmitted disease until AIDS. Syphilis usually begins as a sore at the site of infection but then spreads to the body's internal organs over a period of years. Because its symptoms can vary, it is difficult to identify, and it was often confused with gonorrhea, another venereal disease but with somewhat less serious effects.

Ehrlich had worked at Koch's research institute, where he developed chemical stains and dyes. His guiding principle was that biological activities are determined by specific chemical affinities and are quantitatively measurable. This principle opened the possibility of chemical cures for disease. In the 1890s, Ehrlich worked on the production of immunity in animals, not with microbes but with toxic plant proteins such as ricin, developed from the castor plant. He found that when he administered small doses of the poison and gradually increased them, the animal's body would produce specific antibodies against the poison. This is one principle by which disease immunities are developed, as we saw with diphtheria and tetanus, but in this case no diseases were involved. Ehrlich noted that this immunity could be transferred to an infant via its mother's milk.

At Koch's Institute for Infectious Diseases, Ehrlich then looked at bacterial toxins in the same way that he had looked at plant toxins, and he played a role in the development of diphtheria antitoxin. In 1899, he moved to a new institute built for him in Frankfurt, where he remained until his death. Germany's many state-run research institutes were critical to the growth of bacteriological science, but they also had another effect: any therapeutic agents discovered, such as antitoxins, remained in the hands of the state rather than in those of a private company.

Ehrlich continued to seek specific chemical affinities between diseases and certain chemicals. His research program intensified in the years 1906–10, as he attempted to discover synthetic chemicals that acted specifically on pathogenic microorganisms. Could he find affinities between specific chemicals and specific pathogenic organisms without harming the rest of the body? Could he find an "internal disinfectant" that acted as well as Lister's external ones? Ehrlich employed language of "magic bullets" and "poisoned arrows," which zeroed in on a "parasite" and barraged it with chemicals. He tested his "magic bullets" on isolated disease organisms, first in vitro, then on live animals, usually mice. He believed that only live animals could truly assess a drug's potency and safety.

Ehrlich employed industrial chemists from the dye industry to make compounds, and he tried several hundred. He found that the dye Trypan red appeared to cure mice infected with trypanosoma, protozoans that caused the tropical disease known as sleeping sickness. But the treated animals soon developed resistance to Trypan red, and Ehrlich turned to compounds of arsenic. He and his chemists synthesized hundreds of compounds to find one that combined the least toxicity with the most effectiveness. Compound number 606 seemed especially promising, and Ehrlich patented it in 1907. It killed all the syphilis spirochetes in the test mice without apparent harm to the animal.

Just as Pasteur had been inundated with requests for his rabies vaccine, so now Ehrlich was overwhelmed with requests for the new syphilis cure. However, he continued to test the drug on animals, and he conducted clinical trials on humans before allowing the drug to be distributed commercially under the name Salvarsan in 1910. The German government, which sponsored Ehrlich's laboratory, held the patent on Salvarsan and closely guarded its formula. During World War I, American scientists duplicated the compound and sold it at much-reduced prices.

Salvarsan was not a perfect drug, despite the years of tests. Syphilis followed a complex pathology of several stages, and Salvarsan was effective only in its early stages. Ehrlich's recommendation of one or two large doses of the drug, rather than a series of smaller doses, led to dramatic cures but also to dramatic and sometimes fatal side effects, caused by the toxicity of the drug itself or by allergic reactions to the massive die-off of spirochetes. Close scientific supervision of the drug's complex manufacturing process was also necessary. Ehrlich found a less toxic compound, number 914, a few years later. Dubbed Neosalvarsan, it became the standard therapy for syphilis until the development of penicillin in the 1940s.

Continued use of Salvarsan and then Neosalvarsan by physicians established a lower effective dose and a longer period of therapy. Doctors correlated the clinical progress of Salvarsan and Neosalvarsan with the Wassermann test for syphilis antibodies, providing a model for the new scientific medicine, which depended on the laboratory for every step of the healing process.

Filth Parties and Poison Squads

Pasteur's treatment of Joseph Meister with his not-quite-ready rabies vaccine amounted to human experimentation in much the same way as smallpox inoculation in the eighteenth century. In both cases, the urgent public need for relief served to override considerations of safety and efficacy. As we saw in chapter 3, the eighteenth-century physician James Lind conducted what has been called the first clinical trial when he tested various remedies for scurvy on ill seamen. The development of the germ theory in the late nineteenth century and its accompanying serums, antitoxins, and vaccines necessitated extensive testing for safety and efficacy, first in animals and then in humans.

Before 1900, hospitals were for the poor and often a place of last resort. Hospital patients provided a captive, largely powerless pool of subjects for possible experimentation. In the antebellum US South, most hospital patients were slaves and were therefore doubly captive. By the early nineteenth century, the practice of therapeutic experimentation in hospitals had become standardized. In 1814, the French physician Auguste Chomel (1788–1858) summarized the rules for such experiments. He distinguished two varieties of experiments, one to determine the action of a therapeutic agent against a known illness, the other to find the specific body function that the therapy affected. Experiments, said Chomel, should aim to cure a patient, not to make him ill. As we saw in chapter 4, Elizabeth Blackwell and Anna Kingsford protested against gynecological surgeries on poor women in hospitals, which they viewed as forms of experimentation. When Paul Ehrlich developed Salvarsan in the spring of 1909, it had been thoroughly tested on animals. For the next year, doctors gave the drug to hospital patients around the world and wrote articles in scientific journals attesting to its efficacy, leading to Salvarsan's market release in December 1910. Some of the articles also noted possible severe side effects; by 1914, more than 100 deaths had been attributed to Salvarsan therapy. Salvarsan was not the first therapy to be tested on hospital patients, and it would not be the last.

Prisoners were the first in England to test smallpox inoculation in the early 1720s. Imprisonment itself could be an experiment. The Pennsylvania System of incarceration, based on isolation, introduced in 1829 at the Eastern State Penitentiary in Philadelphia, is now seen as a wide-scale (and long-term) human experiment on the effects of prolonged solitary confinement. Intended to induce penitence, solitary confinement often led to insanity. Charles Dickens commented in 1842, "I hold this slow and daily tampering with the mysteries of the brain to be immeasurably worse than any torture of the body; and because its ghastly signs and tokens are not so palpable to the eye, . . . and it extorts few cries that human ears can hear."[3] This system endured until the early twentieth century. However, by that time scientists and prison administrators had begun to realize that prison populations provided ample human material for a variety of experiments, and amnesty or a reduced sentence for participation in experiments provided prisoners with powerful incentives to participate.

While the germ theory of disease illuminated the causes of many diseases, it also obscured the causes of nonmicrobial diseases. Pellagra, a disfiguring skin disease, was one of these. In 1915, Joseph Goldberger (1874–1929), a physician with the US Public Health Service, believed that pellagra was caused not by a germ but by a dietary deficiency. Goldberger had already tested this hypothesis through dietary experiments in two Mississippi orphanages and a Georgia mental asylum. With the agreement of Mississippi's governor, he then began a controlled dietary experiment on prison volunteers. The experiment was successful in showing the relation between pellagra and diet, but critics were not convinced. Goldberger turned to himself as a subject. Convinced that pellagra was not contagious, he convened what he called a "filth party." He and his assistant injected blood from a pellagra victim into each other. They swabbed out secretions from the nose and throat of another victim and applied them to their own noses and throats and even gave each other capsules they had made from the scabs from pellagra rashes. Goldberger's wife later joined the party.

Somewhat less repulsive but more dangerous were another series of human experiments. Harvey Washington Wiley (1844–1930), named chief chemist of the US Department of Agriculture in 1880, campaigned vigorously for safe food and accurate labeling. In this period, the United States lacked any regulation of food and drink. Manufacturers regularly adulterated food to give the illusion of freshness and purity. To demonstrate the

Lab rabbit, Wikimedia. Photo by Andrew Skowron, posted by Otwarte Klatki, CC-BY-2.0

5.2 The Draize Test

Although Harvey Wiley emphasized food adulteration, the Pure Food and Drug Act of 1906 also led to new tests for toxicity and effectiveness of drugs. Wiley's Bureau of Chemistry (later renamed the Food and Drug Administration [FDA]) administered the act, and its Drug Division tested the effects of drugs on animals. Manufacturers were not required to perform safety tests until 1938, when concerns about drug safety as well as toxic cosmetics led to passage of the Food, Drug, and Cosmetics Act. The FDA took an active role in toxicity testing and in 1939 hired John H. Draize (1900–1992) to develop tests for skin and eye toxicity. In 1944, he coauthored a paper that established quantitative scoring of various tests for toxicity of skin, mucous membranes, or eyes. Draize developed many tests, but the one that still bears his name measures toxicity to the eyes. The Draize test for ocular toxicity employs albino rabbits, which have relatively large eyes and no tear ducts. The test is undoubtedly painful and has been criticized both on the grounds of cruelty and for giving only a crude measure of toxicity. In the early 1980s, Henry Spira (1927–1998) spearheaded a movement against the use of the Draize test, targeting cosmetics companies. This led to a reduction in the use of the Draize test but not its elimination; it is still used. However, several industries that employed animal testing contributed to the founding in 1981 of the Johns Hopkins Center for Alternatives to Animal Testing, which actively develops alternative techniques.

noxiousness of food additives, Wiley recruited young male volunteers to participate in "hygienic table trials." In the US Department of Agriculture's cafeteria, they ate meals laced with borax, saltpeter, or formaldehyde. Increasing quantities of the adulterants were added until the men became ill. A reporter dubbed the young men "the poison squad." They became celebrated in the press and in popular culture, and together with Upton Sinclair's novel *The Jungle* (published serially in 1905), the table trials led to immense public pressure for new protections. Congress passed the Pure Food and Drug Act in 1906.

Probably the best-known example of human experimentation in this era took place in 1900. Yellow fever was a scourge of tropical climates and afflicted many soldiers in the Spanish-American War in 1898. The Cuban physician Carlos Finlay (1833–1915) had hypothesized in 1881 that mosquitoes spread the disease, but he could not prove it. His work inspired Walter Reed (1851–1902), head of the US Yellow Fever Commission in Cuba, and his fellow physicians Aristides Agramonte (1868–1931), Jesse Lazear (1866–1900), and James Carroll (1854–1907) to test this new theory on themselves before proceeding to a controlled trial with soldier volunteers. Reed argued that the lack of animal models for yellow fever necessitated human trials. However, he soon returned to Washington. Agramonte had already contracted yellow fever and was considered to be immune. Lazear and Carroll allowed themselves to be bitten by mosquitoes, and both fell ill with yellow fever. Carroll survived, but Lazear died. Subsequent trials on human volunteers included detailed contracts outlining the risks. All of them survived, although three died in a later experiment, and over the next thirty years, five scientists died of laboratory-contracted yellow fever.

Antibiotics and the Decline of Antivivisection

Despite the success of Salvarsan, few comparable chemical drugs appeared for nearly twenty years after Ehrlich's death in 1915. Bacterial infections remained without remedy. Diphtheria antitoxin, obtained mainly from horses, was a notable success. Antitoxin or serum therapy, as we have seen, was not applicable in all infectious diseases and was of limited use as a preventive. In 1907, Theobald Smith experimented with guinea pigs using a mixture of diphtheria toxin and antitoxin, which appeared to confer long-lasting immunity. By 1913 Emil von Behring had developed a successful diphtheria vaccine. Coupled with the Schick test for diphtheria antibodies (invented by Bela Schick [1877–1967] in 1913), the toxin-antitoxin vaccine, even more than Salvarsan, represented the power of scientific medicine. Diphtheria was a disease of innocent childhood, while syphilis carried the taint of illicit behavior. New York City's health department began to administer the Schick test to all schoolchildren in 1919 and immunized those who tested positive. Rates of infection and mortality plummeted in the 1920s.

Other antitoxins offered mixed results. While tetanus antitoxin was effective as a cure, it sometimes caused fatal allergic reactions. The tetanus toxin was so powerful that a toxin-antitoxin vaccine like that for diphtheria was not feasible. Cholera antitoxin proved to be ineffective, and the decline of cholera owed mainly to improvements in hygiene. Attempts at developing other vaccines had similarly mixed results. A killed-bacillus typhoid vaccine was effective, but early vaccines for pneumonia and meningitis were not. Microbiology was still an imperfect science, and much remained unknown about the actions of microorganisms. At the Pasteur Institute in Paris, at Koch's and Ehrlich's institutes in Germany, and at universities and research centers around the world, mice, rats, guinea pigs, dogs, cats, horses, and other animals continued to be sacrificed to provide another piece of the puzzle of human and animal disease.

Nonetheless, the successes of microbiology had an enormous public impact. Science, it was clear, held the power to cure all human diseases, if not immediately then soon. Two popular American books from the 1920s fostered the image of the heroic scientist. In *Arrowsmith* (1925), the well-known novelist Sinclair Lewis (1885–1951) presented the story of Martin Arrowsmith, a midwestern boy who grew up to become a bacteriologist. Although Arrowsmith's devotion to bacteriological research caused him to neglect family and fortune, there was no doubt of his heroic stature. Lewis vividly depicted the lonely and sometimes tedious life of a laboratory scientist, and his McGurk Institute was closely modeled on the real Rockefeller Institute in New York, founded by the oil baron John D. Rockefeller in 1901 as a privately endowed research center devoted to microbiological research. Lewis gained insights into the scientist's life from his friend Paul de Kruif (1890–1972), a former Rockefeller Institute scientist turned journalist. De Kruif's bestselling *Microbe Hunters* (1926) presented the search for microbial agents as a series of highly dramatic stories, written in a racy, colloquial style. While Lewis and de Kruif humanized the scientist, they also fed the public's hope that these ordinary but dedicated men would inevitably find the answers they sought.

In the next decade, as antibacterial drugs finally began to emerge from laboratories, Hollywood joined in the glorification of the scientist. *Arrowsmith* became a successful movie in 1931. In the 1936 blockbuster *The Story of Louis Pasteur*, the actor Paul Muni won an Oscar for his performance as the hero battling the forces of ignorance. The chapter in de Kruif's *Microbe Hunters* on Walter Reed and yellow fever research was the basis for the film *Yellow Jack* (1938). A few years later, Edward G. Robinson, better known for his mobster roles, looked appropriately intense as Paul Ehrlich in *Dr. Ehrlich's Magic Bullet.* In all these movies, the experimental animals on whom discoveries relied definitely took second billing.

The synthesis of what became known as the sulfa drugs in the 1930s was the first real breakthrough in the struggle against bacterial infections. By that time, some scientists had begun to believe that bacterial infections were somehow fundamentally different from those caused by spirochetes or protozoa and could not be chemically destroyed without also destroying the host—the infected human or animal. The German chemical giant I. G. Farben, a prominent maker of dyes, hired Gerhardt Domagk (1895–1964) in the late 1920s to direct research in bacteriology. In 1932 Domagk found a dye that cured certain forms of streptococcal infection in mice, but he did not announce his discovery, which he called Prontosil, until 1935. A group of French scientists at the Pasteur Institute read Domagk's report and synthesized the drug themselves; then they went further and found the active principle, which they called sulfanilamide. Both French and British scientists suspected that Domagk had isolated sulfanilamide by 1935 but had not announced it because sulfanilamide had already been synthesized in 1908 (although its antibacterial qualities had not then been recognized) and therefore could not be patented by his employer, I. G. Farben. Domagk's 1935 report was vague, and requests made by other scientists for samples of the drug to test were stonewalled for some time. By the late 1930s, however, clinical trials of Prontosil and sulfanilamide were so promising that other scientists sought to develop related compounds that might be of use in other bacterial infections such as pneumonia. The commercial success such a drug might enjoy was a motivating factor for several drug companies, and a group of drugs known as the sulfonamides, or sulfa drugs, began to appear.

5.3 Magic Bullets

Paul de Kruif's 1926 bestselling book *Microbe Hunters* spawned a number of other popular books and movies about medical research. The following passage conveys de Kruif's compelling style:

But Roux had got his start. With this silly experiment as an uncertain flashlight, he went tripping and stumbling through the thickets, he bent his sallow bearded face (sometimes it was like the face of some unearthly bird of prey) over a precise long series of tests. Then suddenly he was out in the open. Presently, it was not more than two months later, he hit on the reason his poison had been so weak before—he simply hadn't left his germ-filled bottles in the incubator for long enough; there hadn't been time enough for them really to get down to work to make their deadly stuff. So, instead of four days, he left the microbes stewing at body temperature in their soup for forty-two days, and when he ran that brew through the filter—presto! With bright eyes he watched unbelievably tiny amounts of it do dreadful things to his animals—he couldn't seem to cut down the dose to an amount small enough to keep it from doing sad damage to his guinea-pigs. Exultant he watched feeble drops of it do away with rabbits, murder sheep, lay large dogs low. He played with this fatal fluid; he dried it; he tried to get at the chemistry of it (but failed); he got out a very concentrated essence of it though, and weighed it, and made long calculations.

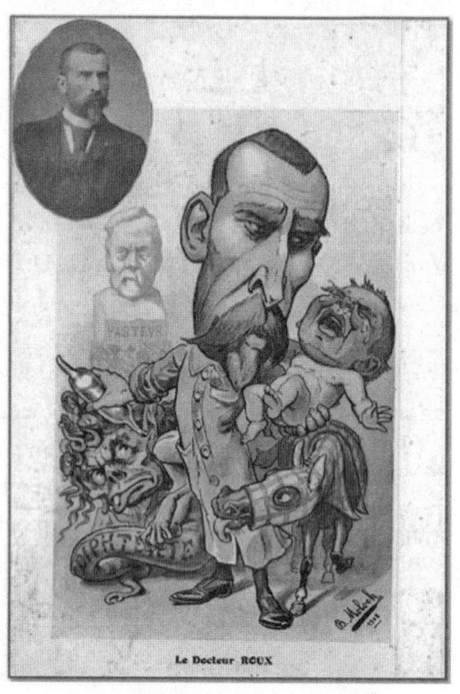

Caricature of Émile Roux, published in the journal *Chanteclair*, 1910. https://commons.wikimedia.org/wiki/File:Dr._Emile_Roux.jpg

One ounce of that purified stuff was enough to kill six hundred thousand guinea pigs—or seventy-five thousand large dogs! And the bodies of those guinea-pigs who had got a six hundred thousandth of an ounce of this pure toxin—the tissues of those bodies looked like the sad tissues of a baby dead of diphtheria.

■ Quote from Paul de Kruif, *Microbe Hunters* (New York: Harcourt, Brace, 1926), 192.

Particularly before resistant strains of bacteria began to develop, sulfa drugs led to dramatic declines in the death rates for streptococcal infections, bacterial pneumonia, meningitis, and gonorrhea. But they proved to be of little use in local infections, such as those resulting from wounds or surgical operations. American university hospitals, with their close ties between clinical medicine and basic research, played a major role in the testing of sulfa drugs on humans; in this, the German research institute, usually separated from other institutions, was not as effective. Yet medical and popular enthusiasm for the sulfa drugs—labeled by the *New York Times* in 1938 as "the drug which has astounded the medical profession"—also led to some inappropriate and excessive use of the drugs by overeager physicians and patients.[4] There were no established rules for clinical trials.

Sulfa drugs were hardly a cure-all, but their development demonstrated that it was possible to make chemical antibacterial drugs. The London physician Alexander Fleming (1881–1955) had found in 1928 that the *Penicillium* mold destroyed bacterial cultures on a plate of agar-agar gelatin. But he was unable to produce an active antibacterial principle from the mold, and applications to human infection of "mold-juice," which he named penicillin, proved inconclusive. Although it did not harm mice or rabbits, he did not test its curative powers on infected mice. Ten years later, in Oxford, Howard Florey (1898–1968), Ernst Chain (1906–1979), and Norman Heatley (1911–2004) returned to the *Penicillium* mold. By 1940, they had figured out how to produce enough of the purified drug to experiment on eight mice that had been infected with streptococci. The four control mice died; the four mice injected with penicillin did not. Trials on seriously ill patients, although hampered by inadequate quantities of the drug, showed equally impressive results. A new antimicrobial drug had been discovered, one that, together with its near relatives, ultimately changed the face of medicine.

By the late 1940s the conquest of infectious disease seemed near, as more and more antibiotic drugs were developed. Most of these drugs, such as streptomycin and tetracycline, were, as penicillin was, derived from natural substances, including molds and soil bacteria. Among pharmaceutical companies in the United States and abroad, competition was fierce to develop a new miracle drug that could be patented. As a result, thousands of bacteria and fungi were tested on thousands of animals to find just one new antibiological agent. In contrast to Ehrlich's 606 attempts, more than

7,000 soil samples were tested in the late 1940s before a mold producing the drug known as chloramphecticol was discovered. Competition, the push for profits, and public demand all fueled an enormous increase in the scale of research.

The antivivisection movement was still active in most European countries and in the United States, but during these years its influence diminished. Shaw's critique of vivisection in 1913 pointed out many of the weaknesses of the medical men's case, but by the 1920s his opposition to vaccination and his noninterventionist views on medical care—views shared by many antivivisectionists—had begun to seem merely cranky in the light of new therapeutic successes. Although the outcome of the bacteriological revolution was, as we have seen, mixed at best by the 1920s, the antivivisectionists' emphasis on physiological research and on pain began to seem beside the point.

Before World War I, antivivisectionists had gained much public support, registering some notable public relations successes. Books such as Louise Lind-af-Hageby and Leise Schartau's *The Shambles of Science* (1903) revealed routine cruelty in medical classrooms and led to open conflict between medical students and antivivisectionists at the site of a statue of a brown dog in London. But these public relations successes were not translated into legislation, and Britain's Cruelty to Animals Act of 1876 remained the only attempt to regulate experimentation. Meanwhile, scientists had begun to recognize the need for public support. Criticism of the use of animals at the Rockefeller Institute, for example, led to a spirited response by its director of research. The American Medical Association founded a Council for the Defense of Medical Research in 1907, which challenged antivivisectionist claims. Its longtime chair, Walter Cannon (the chair of physiology at Harvard Medical School) formulated a code of conduct for animal laboratories that took much of the wind out of antivivisectionist sails. In the 1920s, a noted medical man claimed that antivivisection was "a lost cause."[5]

While antivivisection had lost some public support by the 1920s, it did not die out as a cause, and supporters continued to agitate for reform, if not abolition, of animal experimentation. In the United States, antivivisectionists focused largely on physiological experiments, particularly those on dogs. The use of pound animals and strays and the fear that kidnapped pets were also used featured in antivivisection campaigns from the 1930s to the 1960s.

The historian Susan Lederer has shown that even when antivivisection was at its lowest ebb, researchers were apprehensive about negative antivivisection publicity. Walter Cannon advised editors of medical journals to "eliminate expressions which are likely to be misunderstood" from articles they published, pointing particularly to articles on animal experiments. He recommended, for example, that researchers always mention when anesthetics had been employed. The editor of the *Journal of Experimental Medicine* from the 1920s through the 1940s developed editorial guidelines emphasizing neutral language (e.g., *animal* rather than *dog*) and euphemistic medical terms that distanced the reader from the act of experimentation. Illustrations might show part of an animal rather than the whole. While this strategy may have minimized the grounds for antivivisectionists' criticisms, it also fostered a culture of detachment from the experimental subject. As Lederer notes, "It is not without irony that the apparent linguistic insensitivity on the part of investigators may itself be a historical product of the bitter controversy with antivivisectionists."[6]

By the end of World War II, the successes of microbiology had gained it broad public support. Yet if science could be viewed as a savior, the explosion of the atomic bomb at Hiroshima in 1945 also revealed only too clearly its opposite role as a destroyer. The scale of scientific research grew immensely after the war. Did ethical questions therefore recede into the background?

6 Polio and Primates

My kindergarten teacher in 1958 had had polio as a child and walked with the help of a heavy metal brace on one leg. I can still hear the clump of her step across the old wood floor of the classroom and the click when she unlocked the brace to sit at the piano and sing with her class. These were not unusual sights or sounds in the 1950s. Everyone knew someone who had polio. Polio seemed to be connected with summer and hot muggy days, its attack as sudden as thunderstorms, and just as ominous. One day you were out running around with your friends. Maybe you got a little overheated, and that night you got sick, with a fever and a sore neck. And maybe the next day you would be in an iron lung, struggling to breathe.

Polio, like smallpox, has disappeared so completely from most of the world that its absence seems normal. The last recorded case of naturally occurring polio in the Western Hemisphere was in Peru in 1991. Along with hundreds of thousands of other children, I was injected with the Salk vaccine in the late 1950s and a few years later lined up at school to get a dose of the Sabin oral vaccine out of a little paper cup. It tasted like sugar water.

The conquest of polio is one of the great triumphs of modern medical science. In the United States, the number of new cases of polio dropped from 20,000 per year in the early 1950s (58,000 in the epidemic year of 1952) to 5,000 in 1960 and virtually none a decade later. In 1988 the World Health Organization set as its goal the certification of the world as polio free by 2005, but war, political and social unrest, and uneven funding have meant that parts of Africa and the Middle East retain pockets of polio. Polio cases overall have declined by 99 percent since 1988. As with most scientific accomplishments, the path to the polio vaccine was not straightforward or inevitable, and the conquest of polio was not without its costs. Competition among scientists to develop the vaccine—much like that in

James Watson's account of the 1953 discovery of DNA, *The Double Helix* (1962)—showed them in a new and unfavorable light.

At least 1 million monkeys, and possibly up to five times that number, were sacrificed to develop the Salk and Sabin vaccines. In the light of new research about primate consciousness, intelligence, and sensibility, their use as experimental tools can no longer be taken for granted. The choice of experimental animal in the first decades of research on polio led to the construction of a seriously flawed model of the disease that retarded its understanding for many years. In addition, the procedures followed in the early human trials of the Salk vaccine in the early 1950s raise questions about informed consent, human experimentation, and statistical design.

The viral disease poliomyelitis in its most severe form attacks the lining of the spinal cord, causing paralysis. Although paralytic polio, even at its height, was an uncommon disease, its incidence among school-age children made it seem more common than it was. Depending on the severity of the attack, as well as measures taken during the attack, the paralysis can be temporary or permanent, can affect one or more limbs with varying severity, and can even paralyze the muscles of the chest and diaphragm, preventing breathing and swallowing. Polio can be deadly, but most patients recover, although some with permanent disabilities. In addition, post-polio syndrome can sometimes affect polio survivors years after their initial illness with muscle weakness and joint degeneration. Flaring to epidemic intensity around the world, in the twentieth century the disease was most prevalent in the United States. In the 1952 epidemic, 3,000 deaths occurred among 58,000 cases. The potential of lifetime disability, with its attendant social costs, made polio a much-feared disease. While it most often struck children—thus its alternative label "infantile paralysis"—it could also affect adults. The New York politician Franklin Roosevelt (1882–1945) was 39 when he contracted polio in 1921, and his election as president of the United States 11 years later gave powerful impetus to polio research.

The poliovirus had existed for some time, but only in the 1890s did it emerge as an epidemic disease with the potential for causing paralysis. The irony of a seemingly new disease emerging at the height of the bacteriological revolution was not lost on scientists or the public. By 1910, scientists in Vienna and New York had established that a contagious virus caused polio and that it could be passed from humans to monkeys. But as Pasteur recognized with rabies, no one knew what a virus was. Unlike bacteria, viruses were invisible under conventional microscopes and so tiny that they passed through ceramic filters. The word *virus* signified an unknown infectious agent, not a specific entity. No one knew how viruses could infect the body or where polio had come from. The new science of bacteriology had, temporarily at least, reached its limits.

Viruses remained invisible to researchers until the development of the electron microscope in 1939. Unlike bacteria and other disease-causing microbes, viruses are not cells. Bacteria have their own DNA and independent metabolisms, and they can grow and reproduce (by division) on their own in a hospitable environment. Bacteria are susceptible to antibiotic drugs, which either kill them or keep them from reproducing. Antibiotics have no effect on viruses. A virus is essentially a string of molecules carrying genetic information that is surrounded by some sort of protective protein covering. A virus cannot reproduce on its own. It is a parasite that must have a host organism—a cell—in order to survive. Once it penetrates a cell, a virus can reproduce, sometimes very rapidly, using the host cell's proteins, by injecting its own DNA (or RNA, in retroviruses) into the host's. While many bacteria can infect several different species, viruses are usually much more species-specific because they can reproduce only in genetically compatible cells. The symptoms of viral infection can be caused by changes in the cells themselves or by the body's reaction to the death of large numbers of infected cells. Over time, bacteria can mutate; that is, their genetic structure can change to the extent that, for example, they are resistant to certain antibiotics. Viruses can mutate too, and much more rapidly than can bacteria since they can rearrange their genetic material immediately to suit their host.

We now know that the answers to the questions polio posed were even more ironic than early researchers suspected. The very public health measures that had helped to curb tuberculosis, cholera, and typhoid unleashed the scourge of polio. The poliovirus is carried in fecal matter and (like cholera) enters the body orally. In the days of open sewers and reeking outhouses, exposure to it at an early age was routine. Polio contracted in infancy often manifested itself merely in minor gastrointestinal symptoms. Many infants may have died of polio, but in an age of high infant mortality one disease was often not distinguished from another. Like many childhood diseases, it was endemic rather than epidemic, widely spread through the population rather than appearing in isolated flareups. Public health measures, based on the pre-germ theory connection between dirt and disease, led to closed sewers, clean water, and cleaner food, which ended routine exposure to and infection by the poliovirus. Children, rich and poor, born after the turn of the twentieth century were never infected and therefore never developed the antibodies that could help them resist a more severe attack of the virus. (Franklin Roosevelt, born in 1882, was an exception: his privileged childhood had limited his exposure to the sources of disease.) As a result, in the twentieth century polio struck the unprotected population with unprecedented virulence.

The US polio epidemic of 1916 was a shock to the nation: over a period of five months, 27,000 people, mostly children, contracted the disease, and 6,000 died. New York City was an epicenter of disease, with almost 9,000 cases and 2,400 deaths. Theories of public health based on cleanliness were put to the test, for unlike cholera and other epidemic diseases, this disease did not discriminate between clean and dirty neighborhoods, rich and poor, urban and rural. Although the germ theory, by establishing impure food and water as carriers of disease microbes, had helped to explain why cleanliness could prevent disease, it could not explain this new plague.

With confidence born of the successes in bacteriology of Pasteur and Koch, a few scientists turned to research on polio. But other than acknowledging that a virus caused polio, scientists in the early twentieth century agreed on little else. Even diagnosis was difficult, since in its earlier stages polio resembled many other diseases. The spinal tap—obtaining a sample of spinal fluid by means of a puncture—began to be widely used in 1916, but it was a difficult procedure and did not lead to a foolproof diagnosis. Few hospitals and fewer doctors had the expertise or equipment to analyze spinal fluid for the presence of a virus undetectable under a microscope.

Initially, polio research followed what had by then become the standard procedures in bacteriological research. Researchers attempted to isolate the disease organism and, following Koch's postulates, to produce the disease experimentally in animals. But there was an important difference between polio research and research on other diseases: unlike most other research at the time, polio research relied heavily upon the use of monkeys because only humans and monkeys, it seemed, could get the disease.

Even though Charles Darwin had established the close evolutionary proximity between humans and other primates and Galen had long ago recognized their anatomical similarities, primates were used infrequently in research before the late nineteenth century. Claude Bernard once used a monkey in an experiment, but he refused to do so again because of its resemblance to humans. Later, in the 1870s, the British physician David Ferrier (1843–1928) undertook studies on the localization of brain functions in monkeys, and antivivisectionists criticized him heavily. Unlike dogs or the favored rats and mice of today's laboratories, monkeys were difficult and expensive to obtain. In 1908, however, Karl Landsteiner (1868–1943) and Erwin Popper (1879–1955) in Vienna accomplished an experimental breakthrough. By injecting material from the spinal cord of a polio victim into the spines of two monkeys, they induced the symptoms of paralytic polio in the animals. Subsequent autopsies revealed polio-like changes in the monkeys' spines, establishing that polio was infectious. Landsteiner and Popper established two important principles that guided future polio research: that monkeys could be used as models for human illness and that the seat of polio was in the spinal cord. Neither of these principles, it turned out, was entirely flawless.

Chief among American polio researchers in this period was Simon Flexner (1863–1946), laboratory director of the Rockefeller Institute for Medical Research. Flexner believed that the laboratory was where a cure for polio would be found, and in response to antivivisectionists, in 1911 he announced that a cure was imminent. Flexner's high status disinclined other researchers to enter the field or, later, to challenge his conclusions; and his focus on the laboratory, like Claude Bernard's, excluded the evidence of practicing physicians. Like Landsteiner, Flexner chose the rhesus macaque (*Macaca mulatta*) as an experimental animal. Because monkeys were not yet bred for research, all of them were imported from southern Asia. Both his choice of subject and his method of inducing the disease into the animal, by means of injection into the brain or spine, had unforeseen consequences.

Because Flexner and his research associate Hideyo Noguchi (1876–1928) could not see the poliovirus, they could detect it only in animals that exhibited symptoms. The only way to induce symptoms in a rhesus macaque was through the brain, and the symptoms produced centered on the brain and the spine. This led Flexner to conclude that polio was a disease of the central nervous system. However, if he had used other primates, his conclusion might have been quite different, for other species, including humans, can develop polio by ingesting the virus. We now know that symptoms of oral ingestion—the way most people contracted polio—centered first on the digestive system and only later, after the virus reached the blood, possibly on the spinal cord. Paralysis, in other words, was the exception and not an inevitable result of the disease. Flexner's insistence on the involvement of the nervous system led him to discount reports of physicians who mentioned stomach upset as a symptom and to view paralytic polio as the norm.

Flexner experimented on polio for several years, and like Pasteur, he passed his viral material through many animals. Because viruses are highly adaptable, Flexner's poliovirus adapted itself perfectly to its host, the rhesus macaque. Unfortunately, this also made his research distinctly less applicable to humans. In Flexner's macaques, polio became a virus that affected only the central nervous system. As far as Flexner and Noguchi could determine, the laboratory animals displayed none of the gastrointestinal symptoms clinicians had sometimes noted in human cases. The adaptation of Flexner's virus, therefore, supported his assumptions.

Owing to Flexner's prominence, physicians and the public adopted his conclusions. Flexner's papers include many letters from distraught parents begging for a cure. His insistence that the poliovirus was introduced in humans only via the nose and mouth, from whence it traveled directly to the brain, led to such remedies as useless nasal sprays. His emphasis on paralysis and other neurological symptoms led physicians to ignore nonparalytic cases, which greatly confused the picture of polio's spread and distribution, leaving the means of contagion unexplained. Because Flexner believed that the blood did little to send the virus throughout the body, for many years physicians and researchers disregarded the blood and its antibodies. One historian has written that Flexner's immense influence in the United States "may have delayed understanding of polio's complex clinical and epidemiological features for at least a generation."[1]

The Race for a Vaccine

Franklin Roosevelt's polio helped to change the perception of polio from a childhood disease to one that could strike anyone at any time. Roosevelt's political prominence, personal charisma, and numerous connections made polio research a national priority. The National Foundation for Infantile

Paralysis, founded with his sponsorship in 1937, established the March of Dimes fund-raising campaigns, which raised millions of dollars for polio treatment and research.

Researchers from the 1920s onward began to dismantle Flexner's interpretation of the disease. While rhesus macaques remained the preferred experimental animals, in part because of their availability (they were readily imported from India), other species of monkeys and chimpanzees whose experience of polio more closely mirrored humans' were also used. Laboratory researchers began to listen to physicians and epidemiologists, who then made connections between exposure to polio (even if it led to no symptoms or only mild ones), antibodies in the blood, and immunity, although there continued to be little understanding of how the immune system worked. Each laboratory began with human polio material that it kept growing in a series of monkeys. But someone soon noticed that a monkey infected with one lab's batch of polio was not necessarily immune from infection from another lab's batch. Apparently, different strains (what we would call *genetic variants*) existed, although no one knew how many. In the late 1940s, a program to "type" the strains established that there were three.

Meanwhile, researchers used many thousands of monkeys. Until the 1930s, virologists could not grow viruses outside a living medium. Therefore, monkeys became test tubes, growing the virus in their spines. Researchers killed the monkeys, ground up their spines, and repeated the process. The population of rhesus macaques in their native India plummeted from an estimated 5–10 million in the 1930s to fewer than 200,000 by the late 1970s. Polio research, especially after World War II, contributed to this decimation, and the United States imported 200,000 rhesus macaques a year in the 1950s for vaccine production and research. India began to restrict their exportation in 1955. Already by the 1930s it was difficult to import enough healthy animals to keep research programs going, so the National Foundation for Infantile Paralysis developed an animal facility at Okatie Farms in South Carolina, which provided animals, including monkeys, for projects funded by the foundation and processed imported animals.

The development of tissue-culture techniques in the 1930s lessened researchers' reliance on animals, although it did not eliminate it. The ability to grow a virus in a sterile medium in a laboratory flask, rather than in an animal's body, opened the possibility for wide-scale production of a polio vaccine, which would otherwise require additional millions of monkeys. Early antibacterial drugs such as sulfonamides lessened the possibility of bacterial contamination of the medium. In 1936, Albert Sabin (1906–1996) and Peter Olitsky (1886–1964) grew the poliovirus in a test tube, using human embryonic nervous tissue as their medium. Their use of nervous tissue indicated that they, following Flexner, believed polio to be a disease of the nervous system. Because attempts to grow the virus in other kinds of tissue failed, many scientists continued to assume that Flexner was right. But as Pasteur's rabies vaccine had demonstrated, a vaccine developed from nerve cells might become neurotrophic and carry the risk of fatal brain diseases and allergic reactions. In addition, while embryonic human tissue was the ideal medium for tissue cultures, it was not easy to come by, and the use of such material, obtained from stillbirths and miscarriages, remains controversial. Monkeys were the inevitable substitute, and work on monkeys continued.

Public pressure to develop a polio vaccine was enormous, and whoever developed the vaccine would undoubtedly become famous. Two researchers in the mid-1930s announced that they had developed a vaccine. Maurice Brodie (1903–1939) of New York University employed virus from monkey spinal cords, which he killed with formalin, a solution of formaldehyde. Killed-bacteria vaccines (such as the typhoid vaccine, developed in the 1890s) had been used for some time, but no one was certain that a killed virus could induce immunity. Brodie and William H. Park (1863–1939), director of the laboratories of the New York Board of Health, tested the vaccine on twenty monkeys, as well as on Brodie and some of his laboratory assistants, none of whom experienced ill effects. Brodie proposed a field test—that is, a test outside the laboratory—on 3,000 children. His rush to proceed with human trials was prompted by his rivalry with the Philadelphia researcher John Kolmer (1886–1962), who was developing a live-virus (but weakened) vaccine. Brodie went ahead and vaccinated 1,600 California children, and later, several thousand others. He was forced to stop when several children developed allergic reactions, probably to the animal tissue. It was not clear whether the vaccine conferred immunity. Kolmer also tested his vaccine on himself and on his two children before vaccinating some 10,000 children. But several of these children developed severe cases of polio, and some died, leading Kolmer to withdraw his vaccine. Each presented his results at a contentious meeting of the American Public Health Association in 1935. The hostile response, particularly to Kolmer, discouraged further research for some time.

Brodie and Kolmer stimulated an ongoing debate about killed- versus live-virus vaccines. Most virologists in the 1930s believed that a successful vaccine had to follow the Jenner model: a weaker version of the disease. The poliovirus was generally attenuated by passes through animals or, later, through tissue cultures, although Kolmer used chemicals to weaken the virus. In all live-virus vaccines, there is a danger (not fully recognized at the time) that the weakened virus will back-mutate into a more virulent form. This could not happen with a killed virus, but scientists believed that a killed virus could not generate lasting immunity. The nature of immunity was still unclear. Peter Olitsky and Herald Cox (1907–1986) tested the Park-Brodie vaccine in monkeys at the Rockefeller Institute and announced that it produced antibodies in the blood but that this did not constitute true immunity. They believed antibodies were products of infection, not evidence of immunity.

The failures of Brodie and Kolmer were much in mind when Salk's killed-virus vaccine began to be tested two decades later. But some of the issues of consent that their tests raised remained unresolved. Brodie and Kolmer insisted that they did not vaccinate anyone without their (or their parents') consent. But what did the parents consent to? How much did they understand of the experimental process? When did field trials stop being experimental and start being therapeutic? How many cases added up to proof?

The quest for a polio vaccine intensified in the years after World War II. The National Foundation for Infantile Paralysis had used Hollywood figures and sophisticated fund-raising techniques from the beginning. In postwar America, the March of Dimes poster children served as constant reminders to the public of the importance of conquering polio, although employing images of disabled children to raise funds could be seen as ethically problematic. A significant research breakthrough had occurred in 1939, when a scientist with the US Public Health Service found a strain of the poliovirus known as the Lansing strain, which would grow in the brains of mice and rats. In 1948, John F. Enders (1897–1985), Thomas H. Weller (1915–2008), and Frederick C. Robbins (1916–2003), in Boston, grew the Lansing strain on human non-nervous tissue, opening the door for production of a safe vaccine. Enders, Robbins, and Weller won the Nobel Prize in Medicine in 1954 for this work.

Although tissue culture lessened the reliance on animals, it did not end it. In fact, large-scale vaccine production in the 1950s increased the number of monkeys used. Enders had used discarded foreskins from circumcised baby boys. But as we have seen, the use of human tissue was controversial, and its supply was limited. Tissue-culture techniques meant, however, that material from one monkey could now produce 200 culture tubes. Previously, a single monkey spine had yielded only three samples of virus.

In 1947, the National Foundation for Infantile Paralysis funded a young researcher, Jonas Salk (1914–1995) of the University of Pittsburgh, to work along with researchers at several other laboratories on identifying the different strains of the poliovirus. At New York University and the University of Michigan, Salk had worked with his mentor Thomas Francis (1900–1969), who had developed a killed-virus vaccine for influenza. The work of Isabel Morgan (1911–1996), Howard A. Howe (1901–1976), and David Bodian (1910–1992) at Johns Hopkins confirmed that there were only three strains of the poliovirus, but this knowledge was hard won in a lengthy process requiring thousands of monkeys. The procedure involved infecting a monkey with a known strain, waiting for it to recover, then introducing an unknown strain. If the monkey got ill again, the unknown strain was different from the known one. But that did not in itself identify the unknown strain, which then was further challenged by another unknown. Complicating the procedure were variations in the virulence of the viruses: how much of the virus did the scientist need to give to the monkey to make it sick but not kill it? Monkeys were expensive, and researchers did not like to lose monkeys unnecessarily. Morgan, Howe, and Bodian published their results in 1949. They named the strains Brunhilde, Lansing, and Leon. Brunhilde was named for one of the team's chimpanzees.

Rhesus monkeys are delicate organisms, easily injured and quick to catch whatever disease is in the air. In his laboratory in Pittsburgh, Salk designed housing for the experimental monkeys. He even donated some surplus animals to a local zoo. But it would probably be a mistake to see Salk as being exceptionally concerned about his animals. Because monkeys were expensive and essential to the experimental process, they had to be treated well, at least in terms of physical well-being.

In order to save monkeys, which would save both money and time, Salk turned to tissue cultures in his virus-typing research. Typing a virus in a test tube took days rather than weeks. Tissue cultures could also be used to develop a vaccine. Salk chose to work on a killed-virus vaccine, which would be safer, but its long-term effectiveness was still unproven. Isabel Morgan had shown in 1948 that a killed-virus vaccine could induce immunity in monkeys. Her article on the vaccine criticized Brodie's techniques, and the failure of his killed-virus vaccine did not deter Morgan or Salk.[2] Brodie had apparently used too much formalin, but there was also a danger of leaving live, toxic virus in such a vaccine. The only way to determine whether a killed virus was truly inactivated was to inject it into monkeys' brains, as Morgan had done. But animals varied in their susceptibility to polio and to particular strains; Morgan found that she could infect rats with one strain but not another and that her vaccine was fully effective against only one strain.

Moreover, Thomas Francis was not certain that his killed-virus influenza vaccine conferred long-lasting immunity. Although the relationship between antibodies and immunity was clearer by 1949 than it had been in the 1930s, even in 1953 Salk could refer to the relationship between antibodies and disease resistance as an "assumption," not a proven fact.[3] Other scientists, including Sabin, Cox, and Hilary Koprowski (1916–2013), worked on live-virus vaccines; Howard Howe experimented with a killed-virus vaccine on chimpanzees. Morgan had largely abandoned polio research after her marriage in 1949.

Competition among researchers to develop a vaccine was fierce, and there was a rush to begin human trials. As early as 1950, Salk proposed to the head of research for the National Foundation for Infantile Paralysis that experiments might be tried on chimpanzees and also on institutionalized children and prison volunteers. But the foundation was unconvinced and suggested that he finish the virus typing they had funded him to do. Koprowski, meanwhile, who worked for a commercial drug company, tested his live-virus vaccine on 20 "feeble-minded" children at an institution in New York State. In 1952, *Life* magazine reported that Howe had tested his vaccine on 6 children, noting that "it must be tested on thousands of animals and on more human subjects to make sure that it is safe as well as effective." The article did not mention that the tested children resided at Rosewood State Hospital in Maryland, an institution for the developmentally disabled.[4] Also in 1952, Salk proceeded with human trials. Having

6.1 Suffer the Little Children

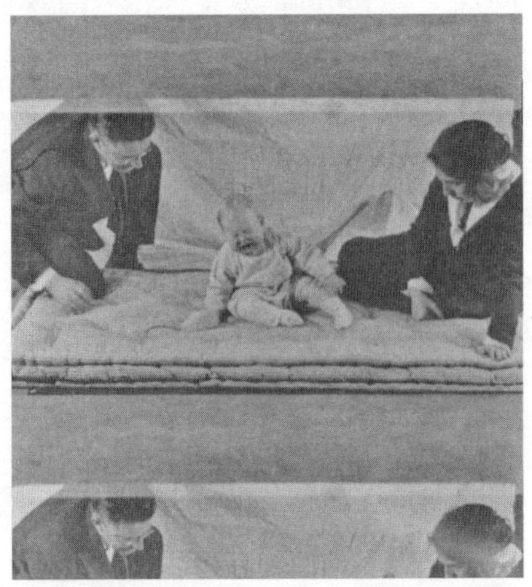

John B. Watson, Rosalie Rayner, and "Little Albert," subject of emotional conditioning to make him fear furry objects. Still from film by Watson and Rayner, 1920. Courtesy of the Drs. Nicholas and Dorothy Cummings Center for the History of Psychology, University of Akron

Smallpox inoculation in the eighteenth century targeted children, and 8-year-old James Phipps was Edward Jenner's first experimental subject for vaccination. While children were highly susceptible to smallpox, their lack of autonomy also made them vulnerable to exploitation. Orphans, as we saw in the 1720s, and other institutionalized children often had no one to speak for them other than institutional authorities. In the 1930s, both Brodie and Kolmer tested their vaccines on children. Of course, polio, or "infantile paralysis," seemed to be largely a disease of childhood, and both researchers claimed that they obtained parental consent for at least some of their test subjects. However, antivivisectionists accused Brodie and William Park of testing their vaccine on orphans. Elizabeth Blackwell and Anna Kingsford had argued that it was but a small step from vivisecting animals to vivisecting poor women in hospitals. Similar arguments had emerged in the second decade of the twentieth century around children, particularly orphans, who were used in several studies of new vaccines and diagnostic tests. Among these were Hideyo Noguchi's 1911 trials of a new diagnostic serum for syphilis called luetin. His test subjects included orphans, hospital patients, and asylum residents. Antivivisectionists claimed that such tests proved their premise that animal experimentation led inevitably to experimenting on humans, but Noguchi's actions also stimulated discussion about consent.

When polio researchers in the 1950s turned to institutionalized children for their early vaccine trials, they had ample precedent. The use of the intellectually disabled, in particular, suggests both a hierarchy between humans and animals (intellectually disabled people were closer to animals) and among human disabilities (intellectual disability was worse than physical disability). In this period, consent was an affair between the experimenter and his subject: there were no institutional review boards, no protocols to be reviewed. After 1946, the American Medical Association required that volunteers consent to experimentation, but this requirement had no legal standing. The legal guardians of institutionalized children signed consent forms; in many cases the guardian was the state.

vaccinated without incident several laboratory workers, himself, and his children, he next tested polio victims at a home for handicapped children. If the vaccine raised the level of blood antibodies in the children, he could claim that it was more effective than natural polio in inducing immunity. Salk, like Howe, then turned to the developmentally disabled, in this case to a Pennsylvania school for such children, as the first site for his vaccine trials. Salk continued to refine and modify his vaccine throughout these small-scale trials.

Few knew of Salk's work in 1952. He announced the results of these trials at a meeting of polio researchers in January 1953, and several participants suggested a much larger field trial. Albert Sabin, supported by John Enders, objected, saying that Salk needed at least 10 more years of research to prove that the killed-virus vaccine was effective and to determine the proper dosage. To Sabin, who pursued the more mainstream science of a live-virus vaccine, Salk's killed-virus, tissue-culture vaccine was too risky. The National Foundation for Infantile Paralysis, eager for a vaccine, promoted the idea of a field trial and enlisted public opinion on its side.

The 1954 Field Trials: Human Experimentation?

Salk published the results of his 1952 trials in the *Journal of the American Medical Association* in March 1953. Even before publication, the National Foundation spread news of the new vaccine to the press and sought support for large-scale field trials. In an interview on national radio, Salk explained his research and attempted to quell rumors that a vaccine would be immediately available. A March 1954 article in *Time* magazine focused on Salk, characterized as "a young man in a hurry."[5] This publicity made Salk highly suspect to his scientific colleagues, to whom publication and evaluation by peers was the only way to assess research. In their eyes, bringing the public into it was a mistake that challenged scientists' autonomy and endangered scientific advance.

Salk steadily continued small-scale field trials in 1953 while also expanding his laboratory. Okatie Farms sent Salk 50 monkeys a week, and his lab also used thousands of mice, as well as chickens. Although tissue cultures meant that Salk used far fewer monkeys than many of his predecessors in polio research, his discovery that monkey kidney tissue produced the best cultures meant that thousands of monkeys were sacrificed. Production of enough vaccine for a large field trial, not to mention enough to inoculate every child in the United States, would take many thousands more.

The National Foundation appointed Thomas Francis to oversee the national field trials, scheduled to take place in 1954. Francis favored a double-blind trial, in which half the children would be injected with a placebo. The subjects would be coded so that no one would know who got the placebo and who got the vaccine until all the results were tabulated. In terms of experimental models and statistical rigor, a double-blind test was by far the best. But to be successful such a trial must also be randomized: the choice of who received the new vaccine and who received the placebo would be entirely random, although children in the same classes and of the same age had to be matched against one another, since geographical variation in the natural incidence of polio was high. The statistical task was daunting, especially since randomized vaccine trials had never been done before. The first randomized and double-blind trial of any sort had been performed in the United Kingdom only five years earlier.

Side by side with the double-blind test in 11 states was an *observed control* test in 33 states, in which the control group was not given a placebo but merely observed. While the randomized placebo model was essential to eliminate the effects of bias on the results and would help to placate Salk's critics, in his words it "would make the humanitarian shudder."[6] However, nonparalytic polio looked a lot like a number of other flu-like diseases, and a doctor could diagnose a vaccinated patient wrongly. The observed control trials were far easier to sell to nervous parents, who disliked not knowing whether their child had received the vaccine. The fear of polio was so great that parents overwhelmingly agreed that their children should be "polio pioneers" and participate in the trials. The consent forms they filled out did not employ the word *consent.* More than 60 percent of parents requested that their children participate in the placebo trials, and almost 70 percent in the observed control trials. Extensive publicity and in-school presentations encouraged parents to understand the benefits of what was about to occur.

Coordinating a field trial of 1.6 million children was an enormous task. It included monitoring several commercial drug manufacturers to ensure vaccine production, which claimed the lives of 4,000 monkeys per month in 1954, in part because each batch of vaccine was tested three times. The

Jonas Salk holding flasks of vaccine. Photo in *The Owl*, the University of Pittsburgh yearbook, 1957. University of Pittsburgh Digital Archives, https://en.wikipedia.org/wiki/Jonas_Salk#/media/File:SalkatPitt.jpg

field trials began in April 1954. The "Polio Pioneers," more than 600,000 first-, second-, and third-graders in several states, received injections of either the Salk vaccine or a colored-water placebo and returned for two more shots. Others had blood samples taken, and more than a million children participated as observed controls. At the end of the trials in June, records of the inoculations were sent to Francis in Michigan for evaluation. In December, Albert Sabin began his own trials, on 12 prison volunteers, of his live-virus vaccine.

On 12 April 1955, Francis announced that the trials of the Salk vaccine had been successful, displaying an effectiveness of 80–90 percent in preventing paralytic polio. On the same day, the US secretary of health, education, and welfare licensed the vaccine for distribution. However, the move to mass production of the vaccine was not without its hurdles. Making the vaccine was a complicated process, and the drug companies assigned to make it had to exercise special care. Only a few weeks after distribution began, some children who had been inoculated with vaccine from Cutter Laboratories came down with polio from a batch of improperly prepared vaccine containing live virus.

Meanwhile, Sabin and Koprowski continued their work on live-virus vaccines. In the late 1950s each conducted field trials—Koprowski in what was then the Belgian Congo and Sabin in Russia. In 1961 Sabin's live-virus oral vaccine (OPV) was licensed for use in the United States, and by 1965 it had mostly supplanted the Salk injected vaccine (IPV). The OPV confers protection by means of the intestines rather than the blood and is therefore more effective in areas of epidemics. In addition, contact with an OPV-inoculated person could provoke immunity by means of infection even in someone who has not been vaccinated. The OPV has the additional advantages of being cheaper to produce and easier to administer.

The Monkey Connection

In June 1999, a federal advisory panel recommended that the OPV be abandoned in favor of the IPV. What caused this dramatic change in policy? It has long been known that live-virus vaccines carry the risk of transmitting the disease they are meant to prevent. Even if the virus does not back-mutate into a more virulent form, the weakened version can in some cases be sufficient to cause disease. Because the OPV contains live virus, it may induce the disease, and a few children every year in the United States contracted polio from OPV. Although the risk is low, one case for every 2.4 million doses, it is real. Because natural polio no longer exists in the United States, elimination of vaccine-caused polio became a reasonable goal.

Other risks associated with the polio vaccine have nothing to do with the poliovirus, but with the use of monkeys to make it. Not only do human viruses such as polio affect monkeys but monkey viruses can also affect humans. In the 1930s a monkey bit a researcher, who then died from a mysterious disease. Albert Sabin and Arthur W. Wright (1894–1976) proved that the death had been caused by a monkey virus, which they named B virus, for their dead colleague. B virus was isolated in the 1940s, and fortunately for Salk, the formalin he used to kill the poliovirus in his vaccine also killed

B virus. But another virus remained in the monkey tissue used to make the vaccine. At the 1953 meeting in which Salk announced his vaccine, some scientists worried about immune reactions to his use of monkey kidney tissue, but no one mentioned that the tissue could carry other viruses. In the early 1960s, scientists discovered a virus known as SV40 (for simian virus number 40) in both polio vaccines, although mostly in the Salk vaccine. From 10 percent to 30 percent of vaccines administered in the years 1955–63 were contaminated. Apparently, no one became ill from SV40, although some studies have linked the virus to certain cancers. However, the evidence, based mainly on statistical studies, remains "inadequate to accept or reject a causal relationship between SV40-containing polio vaccines and cancer," according to the US Institute of Medicine.[7] Nonetheless, the incident raises troubling questions about our use of animals, especially nonhuman primates, to produce vaccines.

Nonhuman primates have been used not only to produce vaccines but also, in a few famous cases, in transplant surgery. The South African surgeon Christiaan Barnard (1922–2001), who perfected human heart transplantation, first tried to transplant the hearts of chimpanzees into humans in the 1970s. The much-publicized case in 1984 of Baby Fae involved transplanting a baboon heart into a human baby; the furor that followed focused mainly on the baby, who died, although some consideration was given to the baboon, who of course also died. In 1992, surgeons transplanted baboon livers into two humans. In none of these cases did the humans survive for more than a few weeks, and the animals died immediately. New technologies, as we will see in the next chapter, may allow custom-made human replacement parts to be grown in animals.

The Right Model

The philosopher of science Richard Burian notes, "Most biologists realize that the choice of organism can greatly affect the outcome of well-defined experiments and can thus have a major impact on the valuation of biological theories."[8] Galen had dissected monkeys because he could not dissect humans, recognizing their close anatomical similarities, and had made assumptions about the human body based on monkey anatomy. As we have seen, Flexner's reliance on monkeys was in part responsible for his flawed model of polio. Nonetheless, monkeys and other nonhuman primates were seldom used in research until the twentieth century. Some researchers, including Claude Bernard, found their resemblance to humans unnerving, and in addition, they were not as readily available in Europe or the United States as dogs or cats or, later, rats or mice. The English physician Edward Tyson (1651–1708) was the first to describe what he called an "ourang-outang" or "pygmie," actually a chimpanzee, in 1699. Tyson wrote that "notwithstanding our *Pygmie* does so much resemble a *Man* in many of its Parts, more than any of the *Ape-kind*," it was nonetheless a "*Brute-Animal*."[9]

Pasteur employed monkeys to attenuate the rabies virus before he turned to dogs and rabbits. In the 1890s, Koch's assistant Richard Pfeiffer (1858–1945) isolated what he believed was a bacterium that caused influenza; he injected many animals with the bacterium but only produced flu-like symptoms in apes. When it became clear during the 1918–19 influenza pandemic that the causative agent was what was known as a "filterable virus," researchers turned to monkeys. Popper and Landsteiner had already used monkeys to demonstrate that polio was contagious. Disease research relied upon many different species of nonhuman primates, but the most common were several varieties of old-world monkeys of the order Cercopithecidae, which included rhesus macaques, cynomolgus macaques (*Macaca fascicularis*), and African green monkeys (*Cercopithecus aethiops*).

Their similarity to humans also made nonhuman primates the preferred experimental model for exploring brain function and behavior. In the nineteenth century, David Ferrier electrically stimulated the brains of macaques, allowing him to assign motor functions to particular areas of the brain. Surgeons then employed Ferrier's neurological map to find the site of a tumor or lesion in a human brain by observing which functions were impaired. Antivivisectionists strongly opposed Ferrier's work with monkeys, and he was unsuccessfully prosecuted for unlicensed surgery on animals in 1881 under the terms of the 1876 Cruelty to Animals Act. In his antivivisection novel *Heart and Science* (1883) the novelist Wilkie Collins (1824–1889) modeled the vivisecting physician on Ferrier.

Ferrier's work concerned the physical structure of the brain, but animals also began to be used to investigate behavior. The Russian physiologist Ivan Pavlov (1849–1936) discovered in 1901 what became known as *conditioned reflex*. Dogs would salivate when presented with food; Pavlov introduced a sound when food was presented, and the dogs would eventually associate the sound with food and salivate in response to the sound. This was the beginning of comparative psychology, which studies animal behavior to understand humans. The American psychologist B. F. Skinner (1904–1990) established an experimental method he called *operant conditioning* based on work with rats and pigeons in a controlled environment (the "Skinner box"). The animals responded to a stimulus, learning particular behaviors to elicit a reward (Skinner thought pigeons could be trained to guide missiles from the nosecone by pecking at a target). The conditioning resulted from both positive and negative reinforcement. But neither Skinner nor Pavlov before him believed that animals employed any form of intelligence or reasoning in learning; the idea that animals exercised rational choice rather than instinct, or that they experienced emotion, was simply nonsensical to them. Skinner's interpretation of the theory of behaviorism, as applied to humans, emphasized external environmental influences, rather than internal mental processes, as the causes of behavior.

By contrast, Skinner's contemporary the University of Wisconsin researcher Harry Harlow (1905–1981) believed that animals, particularly nonhuman primates, had both intelligence and complex emotional lives. Harlow argued that nonhuman primates were capable of complex problem solving, forming concepts, and using tools. The driving mechanism of learning was not simply the satisfaction of earning the reward, the food at the end of the maze; curiosity and solving the problem were themselves motivations. Harlow's continuing study of brain function suggests that he sought a physiological basis for behavior, but he never fully worked this out. In contrast to prevailing behaviorist theories, which asserted the dominant influence of the environment on behavior, Harlow argued that some behaviors were innate rather than learned.

Harlow experimented mainly on monkeys from the 1930s to the 1970s. His studies of the mother-infant bond made him one of the most influential, and most controversial, figures in modern science. With such a favorable view of primate intelligence, Harlow should, it seems, have been a modern hero rather than a villain. During the period when Harlow conducted his research on maternal deprivation, monkeys were subjected in the name of science to radiation exposure, space flight, blunt impact trauma, vertigo, gunshot wounds, and many diseases. But Harlow's research, which explored one of the most basic of relationships, eventually made him a touchstone for the new animal rights movement. Like Magendie in the nineteenth century, Harlow became a model of the uncaring scientist.

Harlow began his work with rhesus macaques in the 1930s. Eventually, he set up one of the first breeding colonies of laboratory monkeys. In some of his intelligence experiments, he would take infant monkeys from their mothers at birth in order to chart their learning process. Although the babies thrived physically, they showed strong behavioral anomalies, and by the 1950s Harlow turned to the mother-infant bond as the subject of his research. He began to build surrogate mothers: first soft, cuddly ones but later cold, hard, vicious ones. He found that babies preferred soft "mothers" to hard wire ones. Even when the wire "mothers" provided all the food, the baby macaques preferred comfort. But given the choice between a cold, wire "mother" and no mother at all, the babies chose any mother, even one that blasted it with cold air or stuck it with spikes. Further experimental variations showed that neonatal monkeys preferred warm wire mothers to ones that were soft but cold.

Plainly, infant macaques craved contact with a mother. What would happen if they were brought up without it? Infant macaques brought up in isolation were fearful and nonsocial, in contrast to the naturally social behavior of macaques. They did not know how to function sexually, and if females became mothers (which Harlow accomplished by tying females down so that males could impregnate them), they were terrible mothers. Mothers who had been raised in isolation had no idea how to nurture. They did not respond to their infants' discomfort and even hurt them.

Harlow pushed his research further and further, with multiple variations on the theme of isolation—from mothers, from one another, and eventually from any vestige of a supporting environment—producing monkeys with more and more abnormal behaviors. In this context, the ultimate experimental situation was what Harlow called "the pit of despair." This was a wedge-shaped metal chamber, deep enough that the monkey saw only the hands of a caretaker, slippery enough that climbing out and basic movement were all but impossible. The pits of despair created monkeys incapable of acting like monkeys, monkeys that were permanently psychotic, so demoralized that they simply huddled, day after day, in the bottom of the pit. Harlow suffered from depression himself, and the image of being sunk in a well of depression may have influenced this experimental model, although in this case Harlow seems to have confused being depressed with the forces that cause depression.

Harlow made significant contributions to psychological research and to child psychology. Early-twentieth-century manuals of child rearing warned

6.2 The Rearing of Infants

Harry Harlow concluded his presidential address to the American Psychological Association in 1958, in which he described his maternal-deprivation experiments, as follows:

The socioeconomic demands of the present and the threatened socioeconomic demands of the future have led the American woman to displace, or threaten to displace, the American man in science and industry. If this process continues, the problem of proper child-rearing practices faces us with startling clarity. It is cheering in view of this trend to realize that the American male is physically endowed with all the really essential equipment to compete with the American female on equal terms in one essential activity: the rearing of infants. . . . But whatever course history may take, it is comforting to know that we are now in contact with the nature of love.

An infant monkey hugs a terrycloth surrogate.
Harry F. Harlow, "The Nature of Love," *American Psychologist* 13, no. 12 (1958): 673–85

■ Quote from Harry F. Harlow, "The Nature of Love," *American Psychologist* 13, no. 12 (1958): 685.

against excessive contact between mother and child for fear of "spoiling" the child. The early behaviorist psychologist John B. Watson (1878–1958), famous for his work with "Little Albert" (see first sidebar above), declared that "mother love is a dangerous instrument" in his influential *Psychological Care of Infant and Child* (1928).[10] Even Dr. Benjamin Spock (1903–1998), whose bestselling handbook on baby and child care first appeared in 1946, warned against coddling in early editions. Harlow was in part inspired by the research of John Bowlby (1907–1990) on maternal deprivation in early childhood among evacuated British children during World War II. After Harlow's research, no one could doubt the critical importance of close mother-infant contact in the early months of life. Subsequent research has detailed physiological changes, including the production of particular growth hormones, in baby monkeys who have maternal contact compared with babies who do not.

But animal activists, and even some researchers, believe that Harlow at some point crossed a line between curiosity and cruelty and ultimately between good science and bad science. But where was the line? Harlow's flippant language in his reports often makes the reader cringe: not only were his isolation chambers "pits of despair" but he referred to the "rape rack" he used to immobilize isolation-reared females for insemination and to "hot mamas." He displayed no concern in the face of the cruelest experiments, although the impact of the experiments was precisely in their emotional power. Was it his penchant for publicity? Which experiments crossed the line? A laboratory inspector criticized Harlow's "pits of despair" as inhumane, but his work continued to be funded and published in respected journals, and colleagues and students at the time—whatever they might have felt later—did not protest. His prominence as a researcher, which included election to the National Academy of Sciences and the American Philosophical Society as well as the presidency of the American Psychological Association, shielded Harlow from criticism. And what of his other isolation experiments? Why were the pits of despair worse?

Perhaps the cumulative effects of Harlow's experiments are what horrify many who look back on them today. One of his former students, John Gluck, has analyzed some of the reasons why Harlow's work was not criticized by those closest to it in its day, and this analysis is relevant to many other laboratory situations. For his students, Harlow modeled research as a war with nature, a constant struggle, and as in war, normal rules of morality were suspended. There are many examples, from Galen to the present, of military metaphors used in scientific descriptions of research. Gluck also points to the division of labor in a modern laboratory, which diffuses responsibility for individual animals among several researchers and students. Rarely is a single animal the responsibility of a single person; more likely, the animal is simply one of many sources of data for the next publication. As we saw in chapter 5, the growth in size of the laboratory from the small basement room of Magendie to the massive, quasi-industrial operation of Ehrlich had as one effect the distancing of the researcher from the research subject. The sociologist Arnold Arluke further notes, in his 1992 essay "Trapped in a Guilt Cage," that uneasiness or guilt feelings among laboratory workers are simply acknowledged, and he describes coping mechanisms that lab technicians and caretakers develop in their day-to-day work with experimental animals.[11]

One argument that has been advanced against Harlow's research is that it did not provide a good model for human behavior. Monkeys are not humans, and different species of monkeys act differently. While rhesus macaques react radically to isolation, other monkey species, even other macaques, exhibit less severe behavioral changes. Maternal deprivation is important to rhesus macaques, but in other species deprivation of the father is more damaging. This criticism strikes at the heart of the research enterprise and may never be fully answered. In psychology, as in Claude Bernard's quest for the laws of general physiology, the discovery of universals remains elusive. For every animal model shown to be inadequate, there are others that work very well indeed.

The case of thalidomide is often presented as a classic example of an inadequate animal model. Thalidomide was a sedative that began to be marketed to pregnant women in Europe in 1957 for insomnia and morning sickness after animal testing on rats and rabbits had shown no ill effects. The dangers to fetuses of mothers' ingestion of drugs (or alcohol or nicotine) were not well understood in the 1950s. But when taken by pregnant human females, thalidomide caused extensive and severe birth defects. It caused the same defects in other primates. Was this a question of the wrong animal model or of not enough testing? The 1938 Food, Drug, and Cosmetics Act in the United States trusted manufacturers to ensure that drugs were safe and allowed new drugs to be distributed while clinical trials were taking place. The US pharmaceutical firm Richardson-Merrell applied for FDA approval of thalidomide in 1960, after samples had already been sent to thousands of doctors to distribute to their pregnant patients as a clinical trial. Dr. Frances Oldham Kelsey (1914–2015), a drug reviewer at the Food and Drug Administration, refused the application, citing inconsistent testing data, and resisted heavy pressure to approve over the next year. West Germany withdrew thalidomide from the market in 1961 after numerous reports of birth defects and other side effects, and the FDA never approved it. Of the few American women who received it from their physicians as an experimental drug, 17 gave birth to children with thalidomide-induced birth defects, compared with some 10,000 worldwide. In a further twist, thalidomide recently has been shown to be effective in a wide variety of disorders, including skin diseases, some immunologic disorders, and even some forms of cancer.

Animal Rights and Animal Liberation

Like the nineteenth-century antivivisection movement, the animal rights movement arose in English-speaking countries. A founding document was *Animal Liberation* (1975), by the Australian philosopher Peter Singer (b. 1946). Singer presented animal liberation as "the obvious next step" following the liberation movements of the 1960s, employing the term *speciesism* (coined by the animal activist Richard Ryder a few years earlier) to convey parallels with racism and sexism and contrasting animal liberation to traditional animal welfare groups.

For Singer as for nineteenth-century antivivisectionists the issue was pain. His philosophy of animal liberation drew from Jeremy Bentham's utilitarianism, discussed in chapter 3, arguing that the true measure of moral principle is the ability to experience pain or pleasure, not some abstract notion of rights. Governments and individuals should aim to maximize the happiness of the greatest numbers. Because animals too experience pain and pleasure, their interests have equal status with those of humans. Controversially, he extended this argument of avoiding suffering in his 1979 work *Practical Ethics*, in which he argued that euthanasia of severely disabled infants was an ethical option parents could take.

Animal liberation quickly became a radical and activist movement, leading to widespread protests against animal use in experimentation, product testing, and food production. Groups such as People for the Ethical Treatment of Animals (PETA) lobbied for the end of cosmetics testing on animals and revealed serious abuses in research laboratories, including the infamous Silver Spring monkeys. Others have vandalized or destroyed laboratories, "liberated" research animals, and harassed scientists, leading to widespread polarization between researchers and activists.

By the 1980s, new approaches to animal ethics had emerged. The American philosopher Tom Regan (1938–2017) argued in *The Case for Animal Rights* (1983) that some animals possess attributes that give them the same inherent value, and thus the same rights, as human beings. These attributes include beliefs and desires, memory, and an emotional life that includes feeling pain and pleasure. In *The Rights of Nature* (1989), the environmental

The wolf OR-7, who in 2011 became the first wolf to enter California in almost a century.
Oregon Department of Fisheries and Wildlife

historian Roderick Nash (b. 1939) described an expanding circle of rights in which ideas of who or what possesses fundamental rights of existence have over time encompassed wider and wider circles of human society and finally the nonhuman world. In *The Sexual Politics of Meat* (1990), Carol J. Adams (b. 1951) found common ground between feminists and animal activists.

Nash and Singer believe that we are more enlightened than our forebears. But in the history of many nations, rights have been neither inevitable nor linear in their development or application. Nor is the relationship between animal rights and environmental ethics straightforward. Some environmental philosophers, such as J. Baird Callicott (b. 1941), hold that animal liberation and environmental ethics are profoundly opposing philosophies. Pain is at the center of both Singer's utilitarianism and Regan's rights theory. But pain, argued Callicott in 1980, is irrelevant to ecosystems, in which pain, death, and violence are part of the natural cycle of life. An environmental ethic concerns the good of the entire biota, whose interests supersede those of individuals. Callicott offered this "holistic" philosophy as an alternative to the individualism of the rights and liberation philosophies.

Apes and Ethics

Research on nonhuman primates, particularly the great apes (chimpanzees, gorillas, and orangutans), has demonstrated how close to humans they are and how they diverge both from humans and from one another. In the twentieth century, the development of field as well as laboratory research in ecology and ethology (animal behavior) revealed complex and highly intelligent creatures with intricate social organization, tool use, and even speech, all the salient characteristics that have been employed to distinguish humans from beasts since antiquity. It also revealed the intricate web of ecological relations that allowed such animals to flourish (discussed further in chapter 7).

Jane Goodall (b. 1934) began her studies on chimpanzees in the wild in 1960, and over a thirty-year span she tracked the lives of three generations at Gombe National Park in Tanzania. She discovered a complex social hierarchy and tool use, as well as meat eating (chimpanzees had been thought to be herbivores), warfare, and cannibalism. She even observed a polio epidemic. Goodall's focus on each animal as an individual—she named each chimpanzee—revolutionized primate research and emphasized the similarity of chimpanzees to humans. It countered the longstanding scientific tradition of impersonality and objectivity; a journal editor of an early article of Goodall's demanded that she change *he* and *she* to *it* when referring to chimpanzees. Subsequent research by Dian Fossey (1932–1985) on Rwandan gorillas and Biruté Galdikas (b. 1946) on orangutans in Borneo supported Goodall's observations on the intelligence and complexity of the great apes. In the 1990s, the anthropologist Craig Stanford (b. 1956) observed chimpanzees as "funhouse mirrors of our ancestry," concluding that hunting and meat-eating were the key to human evolutionary success.[12]

Chimpanzees also could learn language. Beatrix Tugendhut Gardner (1933–1995) and her husband, R. Allen Gardner (b. 1930), both trained in animal behavior, began to teach American Sign Language (ASL) to Washoe

(1965–2007), a young chimpanzee, in the late 1960s. Roger Fouts (b. 1943) and Deborah Harris Fouts took up this program in the early 1970s. Skeptics claimed that the animals simply mimicked their keepers, but Washoe soon displayed an extensive vocabulary and taught Loulis, described as her adopted son, to sign as well. Duane Rumbaugh (1929–2017) and Sue Savage-Rumbaugh (b. 1946) used a keyboard with symbols rather than sign language, with similar results. Savage-Rumbaugh later taught two chimpanzees to play computer games and extended her work on language learning to bonobos (also known as pygmy chimpanzees). Rumbaugh explored chimpanzees' ability to learn number relationships.

Some find this evidence of chimpanzee intelligence exhilarating. Others find it depressing. If chimpanzees are so smart and so closely related to humans, should they be confined in zoos and used as experimental animals? In the mid-1990s, a group of animal rights activists, researchers (including Goodall), and others conceived the Great Ape Project. The project aims "to defend the rights of the nonhuman great primates—chimpanzees, gorillas, orangutans and bonobos, our closest relatives in the animal kingdom."[13] Including apes in the category of persons rather than property would guarantee them the basic human rights to life, individual liberty, and freedom from torture. While the organizers acknowledge that many humans do not now possess these rights, they believe that this is not in itself a reason to withhold them from apes, who, they argue, are worthy of them from both scientific and ethical points of view.

Beginning in the 1990s, several Western nations have moved to ban or restrict research on great apes. New Zealand's 1999 Animal Welfare Act restricted experimentation on "nonhuman hominids" to work that benefits the animals or the species. A European Union directive in 2013 similarly restricted experimental use to research "aimed at the preservation of those species" and in cases of life-threatening human conditions. Only Austria has completely banned experimentation on both great apes and gibbons.[14] In 2013, the US National Institutes of Health (NIH), the primary federal agency for biomedical research, announced that it would retire most of the chimpanzees currently used in NIH-funded research, retaining only 50 for possible future research but not breeding more. Two years later, all projects with chimpanzees had been phased out, and NIH announced that it would no longer support any new chimpanzee research, citing a new designation by the US Fish and Wildlife Service of both captive and wild chimpanzees as endangered species under the 1973 Endangered Species Act.[15]

Experimentation on nonhuman primates greatly increased in the twentieth century. In areas such as polio research (and more recently, AIDS and COVID-19 research), researchers argued that nonhuman primates were essential models for certain human diseases. Their similarities to humans in physical and emotional characteristics also made them excellent subjects in psychological research. But these very similarities make the use of these animals ethically problematic. As evidence mounted of their intelligence, language use, and social structures, questions also mounted concerning the ethics of their use.

7 From Nuremberg to CRISPR

New Rules and New Sciences

US scientific research saw explosive growth following World War II and the creation of the National Science Foundation in 1950. Biomedical research involving animals increased and combined with a growing optimism that science could solve global problems. The development of the polio vaccine was just one example. Epidemiologists in the 1960s spoke of a "health transition" from infectious to chronic disease, but the emergence of AIDS soon tempered enthusiasm. As experimental psychology continued to evolve, animal behavior continued to model human behavior in many studies, including some that studied great apes for insights into prehistoric humans. Prior to World War II, the science of ecology, defined by Ernst Haeckel (1834–1919) in 1866 as the relationship between organisms and their environment, drew on Darwin's theory of evolution and came into its own in the 1930s. In addition, the postwar industrialization of agriculture led to targeted research on farm and rangeland animals. At the same time, public concerns revived about animal and human experimentation, owing in part to exposure of serious abuses, and governments enacted new legislation. In the first decades of the twenty-first century, the sequencing of human and animal genomes led to new kinds of research as well as to new moral dilemmas.

Research on Human Subjects and the Rise of Regulation

The trial of doctors before the Nuremberg War Crimes Tribunal in 1946 led to a code of conduct, the first written set of guidelines for human experimentation. The first prerequisite was the informed consent of the subject, which meant that he or she must be mentally competent, uncoerced, and fully aware of the possible consequences of the experiment. The goal of the experiment had to be important enough to justify the risk to the subject and also be unattainable by other means. The experiment had to be conducted by qualified persons and terminated if the investigator judged that its continuance would result in death or permanent harm to the subject. It also had to be terminated if and when the subject desired it.

The Nuremberg Code established a voluntary regime that was revised and supplemented by a succession of international declarations under the auspices of the World Medical Association, starting with the Helsinki Declaration of 1964. However, for a long while the code was merely advisory. The physician Jay Katz (1922–2008), who helped to define the concept of informed consent, argued that at first US researchers did not take the code seriously. Operating on the principle that the Nazi experiments were outliers, researchers in the United States and elsewhere believed that the code was, he wrote, "irrelevant, or at least too restrictive, for the kinds of experiments generally conducted in the civilized world."[1] Only gradually did national governments move to protect human subjects. Scandals revealed in the 1960s and 1970s showed that medical researchers in the United States and the United Kingdom had long been violating the rights of human subjects, demonstrating the need for legislative regulation.

The doctors' trial exposed appalling concentration camp experiments. Medical scientists had taken the lead in elaborating the distinctively German strain of eugenics called *Rassenhygiene* (racial hygiene), which classified certain ethnic groups, along with sexual "deviants" and the disabled, as threats to the nation's health. Medical personnel had implemented the Nazi solution to the problem, which embraced sterilization, euthanasia, and ultimately mass murder. As the historian Robert Proctor has shown, hospital personnel had euthanized some 70,000 of their patients by 1941. He remarks that "doctors were apparently never *ordered* to murder psychiatric patients and handicapped children. They were *empowered* to do so, and fulfilled their task without protest, often on their own initiative."[2]

Such "lives not worth living" were the ultimate "vile bodies," and Nazi scientists viewed these "defective" specimens of humanity as fit material for medical experiments for the good of the Reich. Their research included studies of the effects of high altitude, hypothermia, and long-term starvation; the efficacy of sulfanilamide in treating gunshot wounds; and the effectiveness of other drugs and treatments. Men had their genitals irradiated and were then castrated. Children were decapitated in order to study brain development at different stages of growth. Researchers tested methods of sterilization and euthanasia and used prisoners to train surgical students.

7.1 Eugenics: Improving the Human Race

Carrie and Emma Buck, 1924. Courtesy of M. E. Grenander Department of Special Collections and Archives, University at Albany, SUNY

Francis Galton (1822–1911), Charles Darwin's cousin and an eclectic scientist and mathematician, coined the term *eugenics* in 1883 to denote "the study of the agencies under social control that may improve or impair the racial qualities of future generations, either physically or mentally."

Darwin and Galton agreed that natural selection could only operate on heritable traits. Galton and eugenicists assumed that many diseases and behavioral traits, including alcoholism and intelligence, were passed on through families. In the early twentieth century, "positive eugenics" encouraged superior people to breed, while "negative eugenics" discouraged those at the lower end of society. Social reformers around 1900 viewed eugenics as a scientific approach to social issues, particularly those of modern urban society, along with public health measures supported by the germ theory. But the quest for the ideal human increasingly linked hereditary differences to race, which referred not only to skin color but also to other supposed inherent biological differences. "Irish," "Italian," and "Jew" were among racial categories, as was "Negro," and eugenics could filter out new immigrants as well as other "inferior" types. By the 1920s, the main instrument for improving the social order was legislation to determine who could reproduce and who could not. Several countries and dozens of American states enacted laws to allow the sterilization of mentally ill or cognitively disabled people in public institutions. In 1927, the US Supreme Court declared that Carrie Buck, described as "a feebleminded daughter of a feebleminded mother" as well as the "mother of a feebleminded child," could be sterilized under the terms of a 1924 Virginia sterilization law. Chief Justice Oliver Wendell Holmes Jr. declared, "Three generations of imbeciles are enough."

Although eugenics became associated with Nazi policies of "racial hygiene" (a term employed in the US by 1915) and mass extermination, many eugenic sterilization laws remained on the books into the 1970s, and coerced sterilization occurred in two California prisons as recently as 2010.

■ Quotes from Garland Allen, "The Eugenics Record Office at Cold Spring Harbor, 1910–1940: An Essay in Institutional History," *Osiris*, 2nd ser., 2 (1986): 225–64, at 225, and Nathalie Antonios and Christina Raup, "Buck v. Bell (1927)," in *The Embryo Project Encyclopedia*, accessed 26 August 2020, https://embryo.asu.edu/pages/buck-v-bell-1927.

At Auschwitz the notorious Josef Mengele (1911–1979) experimented on 1,500 sets of twins, many of them children. More than a thousand people died as a direct result of such experiments, and many more were permanently injured, both physically and psychologically.

Both Proctor and the bioethicist Arthur Caplan argue that these experiments were a logical expression of the values of Nazi medical science. Caplan writes, "Mainstream biomedicine in Germany boarded the Nazi bandwagon early, stayed on for the duration of the Nazi regime, and suffered few public second thoughts or doubts about the association even after the collapse of the Reich."[3] Jay Katz, however, argued that Nazi experiments "had their antecedents" and were unique more in their severity than in their "disregard of human sensibilities."[4]

The doctors tried at Nuremberg defended their actions as consistent with the values of science and their duties as scientists. They claimed that their subjects had been volunteers who had been promised freedom in return for their collaboration, or that they had experimented only on people who were doomed to die in any case, or that the experiments had given their victims a way to expiate their crimes. They argued that they had been following orders and that their training as scientists had given them no grounding in ethics that might justify refusing those orders. They asserted that they had acted in defense of a state engaged in total war and that it was justifiable that a few should suffer for the good of the many. They also cited American experiments involving prisoners to justify their own actions. In the absence of written regulations, US prosecutors could only argue that American researchers had followed unwritten rules "by common agreement and practice," which differentiated them from the Nazi doctors.[5] The physician Andrew Ivy, who acted as an advisor for the prosecution, drew up guidelines on experimenting with prisoners that validated US research. In a 1948 article in *Science*, Ivy summarized the Nuremberg Code and stated that prisoners in US experiments had been volunteers. He criticized the Nazis for making animal experimentation illegal while experimenting on humans.

Human experimentation was not new, but the Nazi experiments plumbed unparalleled depths. During the same period, Japanese scientists tested biological and chemical weapons on humans and animals at a secret site in Manchuria, in occupied China, known as Unit 731. While many Nazi experimenters were prosecuted during the Nuremberg Trials, the United States granted immunity to Japanese researchers in exchange for information about biological warfare, and Unit 731 received no mention in the Tokyo war crimes trials. The Soviet Union did prosecute some Japanese scientists in 1949, but all received lenient sentences, probably in exchange for their knowledge.

Less gruesome but equally heinous abuses occurred at the same time in the United States and in other democratic countries. The most notorious example is the Tuskegee Study of Untreated Syphilis in the Negro Male, conducted in Alabama from 1932 to 1972 under the aegis of the US Public Health Service (PHS). In this experiment, 400 black males, all thought to have late-stage latent syphilis, were left untreated for decades in order to study the natural history of the untreated disease. Although a study in Norway had provided data on untreated syphilis, racist assumptions about the greater susceptibility of blacks and fears of the spread of the disease led to the Tuskegee study. Researchers assumed that the men in the study were not infectious, although that did not always turn out to be true. Some of the men had been treated, if minimally, earlier in the study (with the standard treatments of neoarsphenamine and mercury) but were not told that the treatment had stopped, and they agreed to allow their bodies to be autopsied after death. An important incentive was the payment of burial costs.

According to the historian Susan Reverby, "The PHS was actively deceiving the men into thinking that they were being treated," when they only received aspirin and an iron tonic.[6] When penicillin emerged in the 1940s as an effective cure, at least in early stages of the disease, nothing changed in the study. Although there were many points at which the Tuskegee study could have been stopped, it was not. Reports from it continued to appear, and no one involved raised questions of its morality. The study only ended in 1972, following a public outcry. President Bill Clinton issued an official apology to the living survivors in 1997. In the course of her research on John Cutler, one of the Tuskegee physicians, Reverby discovered evidence of a PHS study of sexually transmitted diseases (STDs) in Guatemala from 1946 to 1948. The purpose of the Guatemala study was to test the use of penicillin both as a cure and as prophylaxis, and it involved infecting at least 1,300 people with STDs without their consent. President Barack Obama issued an official apology to the survivors in 2010.

During the 1960s, other evidence demonstrated the extent to which medical researchers considered themselves exempt from the Nuremberg standards. In 1965 Henry K. Beecher (1904–1976), a professor of anesthesiology at Harvard University, alerted the national press to several unethical studies. Despite hostility from his professional peers, he published a short account of 22 of these experiments in the *New England Journal of Medicine* in the following year, omitting all names. Several studies involved withholding from a control group a remedy that was known to be effective. In each case, the control group suffered, with some individuals acquiring severe disease, while in one case several people died of typhoid for lack of the correct medication. Beecher emphasized in his conclusion the importance of informed consent. The 1964 Helsinki Declaration had stated that "clinical research on a human being cannot be undertaken without his free consent after he has been informed."[7]

While Beecher was pursuing the issue in the United States, a physician, Maurice Pappworth (1910–1994), was doing so in Britain. He too encountered indifference and hostility from his peers. In his 1967 book *Human Guinea Pigs*, he suggested outright that there was no essential difference between the practices exposed by Beecher and himself and those for which the Nazi doctors had been condemned. Indeed, one striking feature of these experiments was that many of their subjects were people whom Nazi ideology might have branded as defectives. The typhoid study was undertaken on charity patients. Other experiments exposed by Beecher were, Pappworth noted, perpetrated on senile hospital patients and on "mental defectives or juvenile delinquents who were inmates of a children's center."[8] One study, which continued for several years after he reported it, took place at Willowbrook State School for the Retarded on Staten Island. This experiment, which was approved and funded by the Armed Forces Epidemiological Board, involved deliberately infecting inmates with a mild form of hepatitis. And of course there were the 400 men of the Tuskegee study, which had not yet come to light.

The most common breach of the Nuremberg Code was the failure to obtain informed consent. Typically, the subjects of these experiments received inadequate information or none at all. In other cases, they were not able to give uncoerced consent as required by the code. Pappworth raised the problem of experiments on prisoners. Noting that the Nazi doctors had cited contemporary American experiments in their defense, he questioned whether prisoners serving long sentences or confined in appalling conditions were in a position to give uncoerced consent when offered relief from their plight in return for serving as human subjects. One of Beecher's examples involved a different sort of coercion: exploitation of a mother's grief. A mother volunteered to have melanoma transplanted into her from her dying daughter in order to gain a better understanding of the cancer and possibly produce antibodies to the tumor. The daughter died the day after the transplant, and the mother died of melanoma a year later.

Scientists who conducted radiation experiments on behalf of the US government during the Cold War committed many ethical breaches. Two well-documented instances involved the administration of tiny doses of radioactive material to developmentally disabled children at two Massachusetts facilities. In the case of one experiment, conducted from 1950 to 1953 at the Walter E. Fernald School, experimenters sent two letters, at different times, requesting parental permission to experiment on the children. Neither letter mentioned that the children would be exposed to radiation, and both falsely suggested that the experiments would be beneficial. An investigation more than 40 years later concluded that in this case at least, the children had probably suffered no harm. Nevertheless, the Massachusetts Institute of Technology and the Quaker Oats Company (which had collaborated in doctoring the children's breakfasts) settled a lawsuit out of court for $1.85 million.

By the 1970s, Beecher's and Pappworth's exposés and the revelation of the Tuskegee study had demonstrated that biomedical researchers could not be trusted to adhere to the principles of the Nuremberg Code without the coercive inducement of national regulation. In the United States this took the form of the National Research Act of 1974, which established the regulatory apparatus that still prevails. It mandated written consent for experimentation on human subjects and the establishment of institutional review boards (IRBs) to evaluate proposed experiments. The act also set up a national commission to identify the basic ethical principles that should underlie the conduct of research on human subjects. In 1979 the commission issued the Belmont Report, which laid down three guiding principles for research on human subjects: respect for persons, demonstrated by requiring subjects' fully informed consent; beneficence, defined as an obligation to maximize possible benefits and minimize harm to the subject (as opposed to an earlier emphasis on simply avoiding harm); and justice, defined as a fair distribution of the benefits and burdens of research. The principle of justice dictates that vulnerable groups should not be used as subjects, but it has also been invoked to require that particular groups be included in a research design. In the early 1990s, the definition of a "normal" research subject as a white male was revised to include women and, in some instances, ethnic minorities. The 1993 National Institutes of Health reauthorization bill required that both female and male subjects be used at all stages of clinical research.

Regulation of Research on Animals

In the early twentieth century, antivivisectionists made explicit connections between experimentation on animals and experimentation on humans. George Bernard Shaw, for instance, argued that doctors who experimented on animals would be more likely to experiment on their patients. The Nazi example challenged this connection, however. One of Hitler's first acts when he became chancellor in 1933 was to ban animal vivisection, in keeping with Nazis' expressed reverence for nature (animal experimentation did not in fact cease under the Reich). After World War II the impulse to control human experimentation worked against regulation of animal experimentation. The Nuremberg Code precluded research on human subjects that could be achieved by other means, and it required prior experimentation on animals as a means of reducing the risks to human subjects. Moreover, advances in medical science before World War II had helped to defuse antivivisectionist criticisms of research, and increased popular respect for the medical profession had challenged the old image of the physician-vivisector.

By 1960, as government support for scientific research reached unprecedented heights, the antivivisection movement seemed to be at its nadir. But the intellectual and social turmoil of the 1960s revived and transformed the old antivivisection debates. As we saw in chapter 6, new theories of animal rights and animal liberation emerged, and antivivisectionist activism brought abuses to light, much as Beecher and Pappworth had done in the case of human experimentation. A 1966 exposé in *Life* magazine of so-called puppy mills that raised dogs for laboratories caused an outpouring of popular disgust; the article was titled "Concentration Camps for Dogs." Later that year, President Lyndon B. Johnson signed the first Laboratory Animal Welfare Act (AWA), which mainly regulated the sale and transport of dogs and cats used in research. It included minimal standards for laboratory animal care but explicitly did not "prescribe standards for the han-

7.2 Model Organisms

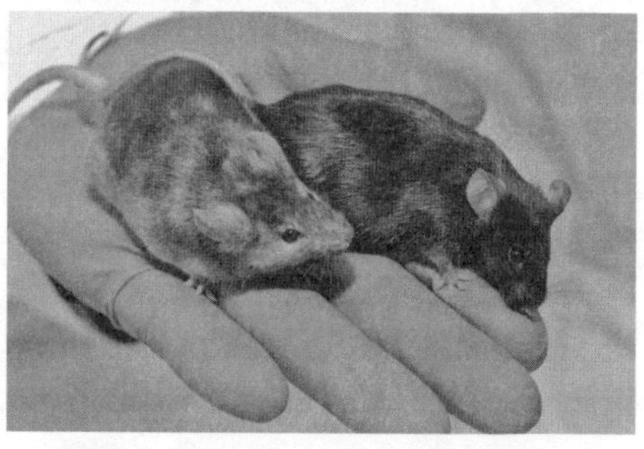

"Knockout" mice. The mouse on the left has a gene that affects hair growth knocked out. Photo by Maggie Bartlett, National Institutes of Health, https://en.wikipedia.org/wiki/Knockout_mouse#/media/File:Knockout_Mice5006-300.jpg

Mice and rats together account for 80–90 percent of all US laboratory animals. Because they are not covered by the 1985 AWA, facilities are not obligated to report the numbers they use, and estimates range from 12 million to 100 million, depending on who is doing the estimating. Since the total number of animals covered by the AWA amounted to 800,000 in 2018, a lower estimate seems more probable. In many ways, they are ideal experimental animals. They are easily tamed and easy to transport and house. Because they can reproduce several times a year, selective breeding can proceed very quickly. The first published scientific work employing albino rats appeared in 1856, and both Koch and Ehrlich had used mice in disease research by 1900. Because of their status as "vermin," mice and rats have received much less attention from antivivisectionists than dogs and cats.

Scientists began to breed rats and mice within closed communities to emphasize certain genetic factors around 1914. Inbreeding assured standardization, which would yield reliable experimental results. They could also be bred to be prone to cancer, or obesity, or hypertension. Over time, scientists acquired vast amounts of data about them, including optimum living standards, and individual animals became points of data. Modifications of the mouse and rat genome have created ever more specialized animals.

In 1999, the Oxford zoologist Manuel Berdoy released his documentary film *The Laboratory Rat: A Natural History*, which followed a group of lab rats released into an outdoor enclosed environment. They quickly reverted to "wild" behavior, including nesting, hunting, and the creation of social hierarchies typical of rats in the wild. It is not clear whether the present generations of genetically modified animals would be able to shed their laboratory life as easily.

dling, care, or treatment of animals during actual research or experimentation."[9] In the rule that resulted from the act, the category "animal" did not include mice, rats, or any non-mammalian animals. Current law explicitly excludes rats, mice, and birds.

Amendments to the AWA in the 1970s established minimal standards of care but little means of enforcement. An exposé in 1981 by PETA co-founder Alex Pacheco of appalling conditions in a monkey research facility in Silver Spring, Maryland, revealed weaknesses in the AWA and turned the focus from dogs and cats to primates. Public pressure led to an AWA amendment in 1985, the Improved Standards for Laboratory Animals Act, and another act applied specifically to NIH-funded research.

The profound changes in scientific practices from the 1960s to the 1980s also owed much to a more pragmatic work, *The Principles of Humane Experimental Technique* (1959), by W. L. M. Russell (1925–2006) and R. L. Burch (1926–1996), two British scientists. Russell and Burch proposed what they called the "three Rs"—replacement, reduction, and refinement—echoing recommendations Marshall Hall had made more than a century earlier. *Replacement* recalled the Nuremberg Code in requiring that animal experimentation be used only when no other means could achieve the goal of the research. Their suggested alternatives included microorganisms and tissue cultures. *Reduction* meant carefully designing experiments, avoiding the trial-and-error approach as far as possible, and employing various techniques to minimize variability and thus reduce the number of subjects needed. *Refinement* applied above all to behavioral studies and entailed the use of lower rather than higher vertebrates and, in general, choosing the appropriate animal for the study.

The influence of the three Rs is evident in the 1985 AWA, which codified higher standards of care and inspection and required the establishment of institutional animal care and use committees (IACUCs), the counterpart of IRBs for human subject research. It imposed a duty to consider the use of alternatives to animal experimentation and set up a central clearinghouse for information on alternatives.

In 1989, Judith Hampson, an expert on animal welfare issues, outlined the general principles of laws regulating animal research:

- to define legitimate purposes for which laboratory animals may be used
- to exert control over allowable levels of pain or other distress
- to provide for inspection of facilities and procedures
- to ensure humane standards of animal husbandry and care
- to ensure public accountability

Hampson argues that no current system has successfully achieved all these goals but that the main elements of an effective system—legislation, a review apparatus, and rules for care and use—exist in many countries.[10] Differing balances among these general aims express national values and specific historical, social, and political circumstances.

Since the 1980s, many nations have adopted some form of legislative control over animal research, mainly guided by the three Rs, although in some countries, such as Canada and Japan, there is no national legislation, and many nations in Africa and Asia do not have regulatory structures. The European Union adopted a resolution in 1986 (last amended in 2010) requiring member nations to regulate animal research, and subsequent legislation by EU states must conform to the resolution. In general, legislation in the European Union is more stringent than in the United States in its restrictions on animal use, as we have seen with primate research. Although there is wide variation among EU nations, the emphasis is on national rather than local enforcement. In another approach, the Australian Code for the Care and Use of Animals for Scientific Purposes, first issued in 1969, provides guidelines for legislation by individual states. Canada and Japan rely on local standards. In Britain, animal research is regulated by the Home Secretary, and inspectors under the Home Secretary's supervision assess and license research projects. Although there are local animal ethics committees, their role is purely advisory.

In the United States, most regulation takes place at the level of the institution, which might be a university, a private company, or a federal laboratory. Committees (either IACUCs or IRBs) composed of scientists and nonscientists review protocols for their adherence to the specified standards. The committees have the power to stop research projects and to require changes in protocols to ensure compliance. However, as Andrea L. Beach and David E. Wright remark with respect to IRBs, despite the increase in regulations and federal supervision, the system remains largely an honor system, and its effectiveness varies among institutions.[11] At some universities, committees meet at least monthly, every member has had an opportunity to review the protocols, and there is serious discussion. In other institutions a committee may serve as little more than a rubber stamp. Ethical considerations are not within their purview (as they are in Australia, for example). The committees' purpose is to ensure compliance with the laws governing research and so to protect the institution, ensuring that external agencies continue to fund research.

Conscientious protocol review requires close attention to complex documents. The terminology can be daunting even for scientists outside the immediate subdiscipline of the protocol; for nonscientist members it can be incomprehensible, leading to rewriting and delay. Mastery of the tangle of laws, regulatory agencies, and acronyms is another lengthy task. In the case of animal experimentation, for instance, institutions receiving federal funding (which includes nearly all US universities) are also subject to rules issued by the National Institutes of Health and administered by the Public Health Service. Unlike the 1985 AWA, which excludes rats, mice, and birds, NIH policy includes rats and mice. In addition, the *Guide for the Care and Use of Laboratory Animals*, developed in 1963 by laboratory veterinarians as a set of self-regulatory guidelines, provides another set of rules. The *Guide*, published by the National Academy of Sciences and now in its eighth edition, presents a set of standards that is generally, but not always, in keeping with federal policy. Voluntary groups such as Public Responsibility in Medicine and Research (PRIM&R) and the Applied Research Ethics National Association (ARENA) hold regular meetings, attended by hundreds of administrators, scientists, and committee members, to provide guidance through the regulatory process.

Regulation is thus necessary but often difficult. Scientists complain of regulatory burdens placed on experimentation and the burden of paperwork required to comply with those regulations. Committee members complain about poorly written protocols and unclear rules. Much depends on the integrity of individual scientists and committee members. A committee's duty is to review the experimental procedures, as set forth in the protocol, to ensure conformity to the law. But often, once the investigator returns to the laboratory, the committee looks over his or her shoulder only infrequently.

Experimenting outside the Lab

Most regulations assume that animal research takes place within a laboratory, where food, housing, and other variables can easily be controlled. However, a significant amount of research occurs outside laboratories, in agricultural settings and in the field. Most research involving agricultural animals such as cows, chickens, pigs, and sheep aims to maximize their value as food. Farmers have been making improvements to animal husbandry for millennia. Selective breeding, conducted largely by trial and error until the twentieth century, preserves desirable characteristics and minimizes those viewed as less valuable. As we saw in chapter 3, the new veterinary schools in eighteenth-century France codified veterinary medicine as a profession concerned with the treatment of animal ills and diseases, especially those of horses and other economically valuable domesticated animals. At the same time, increased recognition of the interplay of animal and human diseases led to a new discipline of comparative medicine, and we can view Jenner's smallpox vaccine as an early manifestation of that discipline. That Pasteur and Koch discovered the germ theory in the context of anthrax, a cattle disease, reveals another side of this complex relationship between animal and human diseases and human valuation of animals. Anthrax was important because it attacked economically valuable animals.

In the United States, two nineteenth-century institutions brought research on agricultural animals into the mainstream of biomedical science: the land-grant universities and the Bureau of Animal Industry (BAI). The land-grant universities were established by the Morrill Act of 1862, which gave each state public lands to be sold or used for profit. These funds would be used to set up at least one college in each state to teach "agriculture and the mechanical arts." The aim of the act was to expand access to higher education in the United States. All land-grant universities have a college of agriculture. The Hatch Act of 1887 added research to the land-grant function and established state agricultural experiment stations, and in 1914 extension agents were added to the system to disseminate the latest scientific knowledge to farmers. In addition, most US colleges of veterinary medicine are attached to a land-grant university.

Agricultural research at the land-grant universities aims to improve the food supply. Diet, breeding, and housing, as well as overall health, are major foci. One example from the early twentieth century concerns eggs. For most of history, chicken eggs were a local and seasonal food. Chickens are sensitive to light and temperature, and in the Northern Hemisphere they lay eggs mainly in the spring. The advent of refrigeration in the late nineteenth century allowed eggs to be preserved for longer periods and to be transported long distances, but cold storage was controversial and led to charges of fraud when old eggs were passed off as freshly laid ones. Extension scientists interested in breeding and heredity turned to chickens. At the Maine Agricultural Experiment Station, for example, scientists attempted to breed a hen that would lay eggs year-round. Dietary experiments also sought to disrupt natural cycles. But the most dramatic changes owed to changes in light, and in the 1920s and 1930s agricultural experiment stations experimented with artificial lighting in henhouses, concluding that hens subjected to round-the-clock lighting produced more eggs than did hens exposed to natural light, particularly in the winter, when egg prices were at their height. This practice of exposing chickens to light twenty-four hours a day quickly exhausted them, but they were deemed expendable in view of the greater profits from their eggs. Since the 1940s, land-grant universities have become powerhouses of biomedical as well as agricultural research, and as such, they exercise close oversight over animal research, even if all the research does not fall under the AWA (which is the case with much research on farm animals, as we will see below).

In the late nineteenth century, the BAI, in association with the veterinarians who began to emerge from the new veterinary colleges, focused its attention on animal diseases. This focus on veterinary medicine contributed to the evolution of veterinarians from "horse doctors" to medical professionals. The BAI was founded amid concerns about the quality of American meat and concurrent outbreaks of cattle diseases. The agency investigated disease outbreaks, conducted animal disease research, and administered laboratories for pathology and bacteriology as well as a field experiment station. The head of the pathology laboratory was Theobald Smith, whom we encountered in chapter 5 for his work on bovine tuberculosis. Smith, a physician, and Fred Kilborne, the director of the BAI experiment station and a veterinarian, published the results of their five-year study of Texas cattle fever in 1893.

Smith and Kilborne combined bacteriological techniques familiar from the work of Pasteur and Koch with experiments on cattle in the field. These experiments included injecting healthy cattle with tissues and fluids from sick cattle, as well as injecting other animals, including sheep, rabbits, pigeons and, in the laboratory, guinea pigs, to discover whether they were susceptible to the disease. Smith and Kilborne concluded that the fever was vector-borne and transmitted by cattle ticks, which they demonstrated by placing tick-infested cattle in a field and then introducing healthy cattle to the field, most of whom contracted the disease.[12] Their work was among the first to establish the role of vectors in the transmission of diseases. It also demonstrated the critical importance of laboratory science to agriculture, and set veterinary medicine on a track parallel to that of human medicine, as a science rather than an art.

The BAI was dismantled in the early 1950s, just as US agriculture was moving from smaller farms to an industrial model. With the help of funding from the US Department of Agriculture and other sources (including industry), land-grant universities, in newly named departments of animal science, aimed to maximize production of agricultural products, as well as to perform basic research on farm animal diseases, behavior, and well-being. In addition, in 1964 Congress designated the creation of the US Meat Animal Research Center (USMARC) in Nebraska. According to its website, USMARC scientists "are developing scientific information and new technology to solve high priority problems for the US beef, sheep, and swine industries. Objectives are to increase efficiency of production while maintaining a lean, high quality, safe product; therefore, the research ultimately benefits the consumer as well as the production and agri-business sectors of animal agriculture."[13]

USMARC "is cooperative with land-grant universities." Its research units include meat safety and quality, nutrition and environmental management, reproduction and genetics, breeding, and animal health. Under the latter category, USMARC has sequenced the genomes of 46 breeds of cattle. Farm animals have an ambiguous status under the Animal Welfare Act. Those used for "traditional production agricultural purposes" are exempt; however, the AWA covers those used in research, although qualifying research includes only practices related to human health.[14] USMARC has an IACUC, and its activities are influenced by the *Guide for the Care and Use of Laboratory Animals* as well as by the separate *Guide for the Care and Use of Animals in Agricultural Research and Teaching*. Each of these is produced by independent non-governmental agencies, and their instructions are advisory rather than mandatory.

USMARC's relative obscurity and lack of oversight has led to serious abuses. A 2015 *New York Times* investigation exposed a wide range of mistreatment of animals, particularly related to reproduction experiments, but also general issues of inhumane treatment and lack of veterinary care. In response, the USDA issued an audit report in 2016 stating that only 7 of the 33 statements in the *Times* exposé were accurate and that the audit had not found evidence of a systemic problem. The audit report did recommend improvements in oversight.[15] Attempts to amend the AWA to include agricultural research have not been successful.

Another kind of field science comes under the umbrella of *ecological science*, which embraces a range of studies, from broad and highly mathematical population ones to intimate investigations of animal behavior. A popular notion of ecological research pictures rugged scientists observing nature in the wild, from a distance. There is some truth to this image, but ecological research can also be as invasive and manipulative as other types of animal research. Much ecological work now occurs in laboratories and with computer models, but it continues to be distinct from other biological sciences in its persistent engagement with field work. Early ecologists tried to capture this distinctiveness in varied titles. Around 1900, Frederic Clements (1874–1945) called himself an "outdoor physiologist," while in the 1930s Victor Shelford (1877–1968) referred to ecology as "scientific natural history."[16]

Shelford, the father of animal ecology, studied animal communities (*Animal Communities in Temperate North America*, 1913), and in experimental studies he combined field and laboratory work (*Laboratory and Field Ecology*, 1929). Shelford's "physiological life history" combined field observation with laboratory experiments that measured physiological responses to environmental variables. Although this project foundered on the number of animals and the complexity of interrelationships in any given ecological community, it introduced methods and concepts that proved to be influential.

We have already seen examples of field work in Jane Goodall's observational studies of chimpanzees. Ethology, the study of animal behavior, gained recognition as a scientific discipline when the zoologists Konrad Lorenz (1903–1989), Nikolaas (Niko) Tinbergen (1907–1988), and Karl von Frisch (1886–1982) were awarded the Nobel Prize in Physiology or Medicine in 1973. Lorenz and Tinbergen combined investigations in the field with investigations in the laboratory and made experimental interventions as well, such as exposing wildfowl to different artificial predators.

Other modern ecological studies involve varying levels of human intervention. Scientists capture wolves and equip them with radio collars to track their movements. A wide variety of animals, including mammals, fish, birds, reptiles, and amphibians, can be implanted with Passive Integrated Transponder (PIT) tags. PIT tags first came into use in the 1980s to track fish movements, such as salmon migration. A PIT tag consists of a microchip, a capacitor, and an antenna encased in a tiny glass tube that can be implanted using a large-gauge needle or surgically, depending on the animal. The tag is then activated via a scanner and provides a unique ID for each individual animal, allowing it to be tracked throughout its lifespan.

As economically valuable species, fish have been subjects of a variety of studies inside and outside laboratories. For example, artificial streams allow experimental manipulations; *electrofishing*, in which the fish are stunned by an electric current and caught in a net, allows measuring, sampling, or other forms of examination. In some cases, nonvalued fish are exterminated from an ecosystem to allow other species to flourish. Not all these activities qualify as experiments, although they do disrupt natural activity in some way. *Lethal sampling*, that is, killing one or more individuals of a targeted species for further analysis in the laboratory, occurs both with birds and with fish. One result of centuries of lethal sampling is the existence of animal skins and taxidermied specimens in museum collections, where they have been valuable resources for DNA analysis. On the other hand, lethal sampling is problematic with endangered species, and perhaps with any species.

Another ecological experiment is *rewilding*. While reintroduction of extant species into places where they have become extinct is a longstanding practice—the reintroduction of wolves, ongoing since the 1980s, is one example—rewilding takes this a step further. In 2005, the ecologist Josh Donlan suggested restoring the lost megafauna of North America by bringing large wild vertebrates from elsewhere to fill their ecological niches. In other words, mammoths, American lions and cheetahs, and the ancient *Camelops*, all of which disappeared at the end of the Pleistocene some 13,000 years ago, could be replaced with analogous species, including elephants, African lions and cheetahs, and camels. Donlan and his coauthors argued that such active management was the only way to preserve and even reinvigorate wilderness areas.

While the "Pleistocene Park" Donlan envisages has not yet come to pass in North America, rewilding efforts have taken hold in widely varying places, including Siberia and New Zealand. The best-known and most controversial of these parks is Oostvaardersplassen in the Netherlands. Beginning in the 1980s, the ecologist Frans Vera aimed to re-create a Pleistocene landscape of grazing animals that could gradually also include forest. In place of extinct wild tarpans and aurochs, Vera substituted Konik horses from Poland (thought to have been descended from tarpans) and Heck cattle, developed by back-breeding in the 1920s to mimic the extinct aurochs. Red deer, more like elk than deer, were added later. Wild animals have since migrated to the site; these include graylag geese, white eagles and many other birds, roe deer, foxes and stoats, and several small rodent species.

A central problem of Oostvaardersplassen is its lack of predators. Before 2018 the park's animals, classed as wild, were not fed or otherwise managed, reflecting a belief that nature could manage itself. Without predators, the site became overpopulated, and in the harsh winter of 2017/18 thousands perished, mostly from starvation. Because Oostvaardersplassen is a fenced reserve (of some 6,000 hectares, almost 15,000 acres), its animals cannot seek other grazing grounds. In 2018, more aggressive management plans led to a reduction in the number of animals—by killing or relocation—and feeding some of them. This experiment has had mixed results.

De-extinction: The Jurassic Ark?

Until the end of the eighteenth century the idea of extinction was inconceivable. Darwin's theory of evolution in the mid-nineteenth century placed extinction at its center as an outcome of natural selection, a one-way process. Michael Crichton's (1942–2008) 1990 novel *Jurassic Park* introduced to the public the idea that scientists could bring back extinct species, proposing that dinosaur DNA found in insects preserved in amber could form the basis for re-creating them. After the birth of the cloned sheep Dolly in 1996 (further discussed below), extinct animals began to get serious attention. In 2000, scientists in India proposed cloning the Indian cheetah, extinct since 1953. Australian scientists turned their attention to the thylacine, or Tasmanian tiger (extinct since 1936), while scientists in Russia considered the Pleistocene woolly mammoth.

Only recently has synthetic biology made de-extinction a possibility. The geneticist George Church (b. 1954) in *Regenesis* (2012) cited the 2003 cloning of the recently extinct Pyrenean ibex, or bucardo, as evidence that extinction is no longer permanent. Critics point out that the cloned bucardo lived for only seven minutes, succumbing to lung malformations that had also affected Dolly, though less severely. In 2015, Church successfully copied woolly mammoth genes into the genome of an Asian elephant with the new CRISPR tool (see below) and opened the door to creating a hybrid creature. Geneticists also see possibilities for improving currently endangered species by introducing disease resistance or diversifying small inbred

Passenger pigeon. Photo by James St. John, Wikimedia, CC attribution 2.5 generic

populations. In such contexts, rewilding takes on a new meaning, and scientists, ethicists, and legal scholars now debate possible consequences of the genetic engineering of the environment. Animal welfare has not been at the forefront of this discussion. For example, little has been said about the 57 goats implanted with engineered bucardo eggs in order to yield a single birth. The old idea of the organism as a machine is here taken to new lengths as the animal is reduced to its genetic code.

Ecologists point out that the environment is constantly changing and that, for example, the niche that passenger pigeons (extinct since 1914) occupied may no longer exist. The major cause of modern extinction is habitat loss, and reintroducing formerly extinct animals will not in itself re-create habitat. Among potential risks are the introduction or spread of diseases, unexpected species interactions, and invasiveness. Some are optimistic that these risks can be overcome. Josh Donlan, who has added de-extinction to his agenda for Pleistocene rewilding, points to successful island eradications of invasive species as an example of humans' ability to manage wild populations. But, he notes, "we are currently better at manipulating genomes than at rewilding landscapes."[17] The ethical implications are numerous. Some contend that de-extinction can undo the wrongs of the past; others believe that it is mere hubris. The legal and social implications, which include patenting organisms, regulation, and funding, are daunting. Moreover, if extinction is viewed as reversible, current protections for endangered species might disappear. Meanwhile, species loss and biodiversity decline continue. Perhaps the genetic tools of de-extinction can mitigate some of these losses, but animals cannot be entirely reduced to cells and genes, and ecological science still has much to learn about and from animals in the wild.

Engineering Life

The development of the science of genetics after World War II revolutionized the use of animals in science and had a major impact on research in human and animal health and reproduction. The historian Nathaniel Comfort writes that "since Galton, scientists interested in human heredity had been dogged by the fact that one could not ethically carry out breeding experiments." The solution, Comfort adds, would be found in "genetics without sex," as the discovery of the structure of DNA—essentially a problem of biophysics—combined with cellular-level studies of heredity.[18] This work began with bacteria but moved on to animal cells, which could readily be cultured for experimental purposes. Human cells were much more difficult. John Enders, Frederick Robbins, and Thomas Weller in the 1940s had successfully cultured human tissue for the first time in their vaccine studies. However, much of the widespread use of human tissue cultures was owing to the cells of Henrietta Lacks (1920–1951), a poor black woman whose biopsy for the cancer that ultimately killed her led to her tumor cells being cultured without her consent, becoming widely used as the HeLa cell line.

A series of discoveries in the 1970s and 1980s placed the gene at the center of both therapy and biology. In the early 1970s, recombinant DNA, which brings together genetic material from different organisms, opened the door to creating genetically modified organisms. Concerns about the ethics and safety of recombinant DNA, including the possibility of human applications, led to a moratorium on its use. In 1975, scientists convened a conference at Asilomar, California, to determine the extent of the existing technology and establish guidelines for its use (or non-use). They lifted the moratorium and established a new NIH Recombinant DNA Advisory Committee. This conference was later viewed as a turning point in the genetic engineering of organisms.

The first genetically modified mouse appeared in 1974. In 1981 the first transgenic mouse, with genes from two different species, was created by injecting a single gene into a newly fertilized egg. The first transgenic rat appeared in 1990. Recombinant technology can also create "knockout" mice and rats, in which a single gene sequence is "knocked out" or "knocked in." Such mutations can be heritable, creating new strains of transgenic animals. However, the ethical implications of such interventions remain poorly defined.

Scientists sequenced the mouse genome in 2002 and that of the BN rat, the most common laboratory breed, in 2004. The rat genome showed not only how far apart rats and mice are but also why the rat has been such a good model for the human body. Knowledge of the genome allowed for more precise genetic manipulation, creating ever more specialized rodents. Researchers can leaf through catalogs of mice and rats and order exactly the right animal tool for research.

The cloning of the sheep Dolly in 1996 revealed the inadequacy of current ethical systems to address new technologies, as commentary at the time focused mainly on the possibility of cloning a human being rather than on the implications of the process itself or its impacts on animals. Somatic cell cloning involves transferring the nucleus of a somatic cell (a cell that is not a sperm or egg cell) into an oocyte (an immature egg cell) from which the nucleus has been removed. The oocyte is then stimulated with an electric current to begin cell division. It was significant that Dolly's donor cell was a somatic cell from a mature sheep rather than a stem cell (an undifferentiated embryonic cell). Creating Dolly showed that it was possible to turn back the developmental clock.

Most biologists viewed Dolly as an interesting but hardly earth-shattering example of reproductive research. Over the previous two decades assisted reproductive technology, including the recovery of oocytes, in vitro fertilization, embryo culture and transplantation, and microinsemination, had led to genetically modified mice and rats, as well as the first "test-tube" human baby, Louise Brown, born in 1978 from an egg fertilized outside the mother's body and then implanted in her womb. Somatic cell nuclear transfer for reproductive cloning remains a difficult procedure with limited success; 277

unsuccessful attempts were made on sheep before Dolly, and the attempt to revive the extinct bucardo required more than 50 attempts. High rates of embryonic and fetal deaths continue to occur. Nonetheless, some 23 mammalian species have been cloned since Dolly, including cats, dogs, and cattle as well as mice and rats. In 2018, Chinese scientists successfully cloned primates for the first time, with the birth of twin long-tailed macaques cloned from fetal cells. Out of 149 embryos created from the fetal cells, 79 survived long enough to be implanted into female monkeys. Of these, four became pregnant; two miscarried, and two carried their fetuses to term. But only the twins survived. In January 2019, Chinese researchers announced the birth of five cloned gene-edited macaques. Unlike most Western countries, China has few restrictions on primate research.

The ability to clone primates inevitably brings up the question of human cloning. Many countries have passed laws that allow cloning of human embryos for research purposes but not for reproduction. After four years of study and debate, the United Nations passed a resolution in 2005 that called upon member states to prohibit both human cloning and genetic engineering, citing the possibility of violating both human and women's rights. But the resolution concluded that member states should also "take into account the pressing global issues," such as AIDS and tuberculosis, in financing medical research.[19] The ambiguity of this resolution seems to leave the door open for therapeutic cloning.

While therapeutic cloning remains a dream for now, the possibility of gene therapies has existed since the 1990s. Even before the discovery of the structure of DNA in 1953, scientists speculated about genetic cures for human genetic diseases; with knowledge of DNA, this could conceivably proceed at a molecular level. The idea of prenatal screening and counseling for genetic diseases emerged in the 1950s, although it was not put into practice until later. In the 1980s, scientists employed new gene-mapping techniques to locate disease-causing genes, including those for cystic fibrosis and Huntington's disease. Cloning the genes then allowed for their intensive study. Attempts at gene therapy were not far behind: for example, could injecting the gene that encodes healthy red blood cells provide a cure for sickle cell disease? Scientists knew little about dosages, risks, or methods. The new apparatus for regulation of human subjects research, established in the United States in the 1979 Belmont Report, tackled the new biotechnologies and their implications for human health. Complications quickly arose. What was the difference between treatment and enhancement? If treating somatic cells was allowed, why shouldn't such treatment also enter the germline and become heritable?

Attempts at gene therapy in the 1990s, some apparently successful, some not, culminated in the death of 18-year-old Jesse Gelsinger in a gene therapy trial at the University of Pennsylvania in September 1999. Gelsinger, who had a rare genetic disease of the liver, volunteered for the experimental procedure. He was injected with a virus vector carrying a corrected gene. In theory, the virus would penetrate his liver cells and insert the corrected gene; however, the virus itself triggered an immune response that led to multiple organ failure.

The resulting investigation by the FDA discovered ludicrously inadequate record keeping by the researchers and other serious lapses. Deaths of experimental animals went unreported, as did serious human side effects. Changes in the experimental design, which should by law have led to the preparation of a revised protocol for review, also were not reported. Most seriously, the researchers were charged with neglecting informed-consent procedures. Jesse Gelsinger was not informed of the death of experimental monkeys from the procedure he was about to undergo. Because of his poor health, he was not actually eligible to participate in the program he had volunteered for. The FDA suspended human research at the Penn Institute for Gene Therapy in January 2000, a move that shocked the public and the research community. The institute was eventually shut down, and gene therapy fell into disrepute for some time. In 2002, a moratorium was declared after a French case in which the retrovirus vector used to insert the gene caused leukemia in a patient.

Less than two years after Jesse Gelsinger's death, in February 2001, the journal *Nature* published a first draft of the human genome. A full map and complete sequence of the human genome followed two years later. The Human Genome Project had been launched in 1988; included within its goals were the mapping and sequencing of the genomes of the most common laboratory organisms, including the mouse (2002) and rat (2004), the fruit fly (*Drosophila melanogaster*, 2000), the *E. coli* bacterium (1997), yeast (1996), and the *C. elegans* roundworm (1998), as well as sea squirts and puffer fish. Since then, the genomes of many other organisms have been sequenced.

Knowledge of the human genome has opened the door to many things, ranging from research in human ancestry (both ancient and modern), to testing for genetic diseases and disabilities, to pinpointing drug responses and developing new drugs. The genes of food animals and plants can be genetically modified to add nutrients or increase the rate of growth, although some of this work, such as the genetically modified salmon known as the AquAdvantage salmon, predates the sequencing of its genome. The AquAdvantage salmon grows year-round rather than seasonally and can thus reach market size in half the time of ordinary salmon. Knowledge of genomes, however, allows such modification to proceed with greater efficiency and precision.

A 2012 article in *Science* magazine introduced what one commentator referred to as "a Swiss army knife of immunity" (and, it turned out, much more than that).[20] The article described a new genome-editing technique known as CRISPR-Cas9, short for "clustered regularly interspaced short palindromic repeats" and "CRISPR-associated protein 9." It employs the Cas-9 enzyme to cut a specific sequence of DNA in a genome and allows for adding and deleting genetic material, which can be customized for a specific function. This technique, developed by Jennifer Doudna (b. 1964) and Emmanuelle Charpentier (b. 1968), is one of many, but its speed and efficiency gained it immediate popularity among scientists. Gene editing can create new mice and rats and make breeding new farm animals and food plants simpler and cheaper. The possibility of eliminating such genetic diseases as the blood disorders hemophilia and sickle cell anemia, as well as other disorders dependent on a single gene, has made CRISPR-Cas9 a subject of massive speculation and (mis-)representations in popular culture. It has been employed in a number of different species, from yeast to mosquitoes to monkeys to pigs. In 2018, a Chinese scientist edited the genome of human embryos. The twin girls were successfully brought to term, becoming the first CRISPR-edited human babies.

While most uses of CRISPR and other gene-editing techniques have been confined to somatic cells, editing of the germline, which would create heritable alterations in the genome, is possible and has been accomplished in plants and animals. The monetary and scientific value of genetically engineered rats is enormous. A recent catalog from Cyagen, one of many providers of animal models for science, lists a group of three CRISPR-Cas9 knockout rats for prices starting at $17,000; more complex knockout models list for as high as $36,000 for three. These would then become the basis for a breeding colony. "Humanized" mice and rats, with human genes, cells, or tissues, are readily available. Pig and cattle heart valves have been transplanted successfully into humans since the 1970s. In January 2022, a heart from a genetically modified pig was transplanted into a human for the first time.

Inevitably, as with cloning, the question of germline editing in humans arose almost immediately. The ethics of human germline editing quickly became a question of how it could be ethical not to employ it for such diseases as Huntington's. I have a friend with Huntington's. She is in her early thirties and bears the genetic markers for early onset of the disease, so she probably has ten years before she starts showing symptoms, first muscle spasms and then increasing dementia. If I asked her whether she would approve of genetic engineering to eliminate Huntington's, I'm pretty sure she would say yes. But most genetic disorders are far less straightforward. Many of them are not diseases. Jennifer Doudna, in her book on her discovery, lists the "diseases for which potential genetic cures have been developed with CRISPR."[21] These include achondroplasia (dwarfism) and congenital deafness, neither of which is a disease, as well as behavioral and cognitive abnormalities. Writers on disability question whether a flawed idea of "normal" humans devalues humans of different abilities, so that a kind of neo-eugenics of eliminating those characteristics deemed abnormal seems like a good idea.

New technologies have given humans unprecedented power to alter animals and ourselves. We can now, or soon will be able to, eliminate certain diseases, create animals to order, and alter the course of evolution. Soon, if enthusiasts are to be believed, we will be able to resurrect extinct animals, manufacture new organs, and eliminate many diseases and disabilities. Who makes these decisions and what values they reflect remain vague. Who determines the value of a human or animal life? Or, perhaps more pertinently, whose voice gets listened to? By making animals into customized tools, even if they are well housed and well cared for, do we deprive them of their autonomy, their ability to be animals? "Human improvement," as the eugenicists of the early twentieth century expressed it, may seem like an unarguable good. But, as we saw over the course of the twentieth century, it also implies that people as they are, with all their diverse characteristics, are not equally valuable and worthy of life.

Conclusion

As noted in the introduction, the COVID-19 pandemic brought new attention to issues surrounding animal and human experimentation.[1] The development of vaccines and therapeutics for this and other diseases, pursued in laboratories around the world, includes experiments with animals and human clinical trials, following procedures similar to those in the development of the polio vaccine that we saw in chapter 6. Scientists readily acknowledge that some animal models are better than others and that mice, the most common experimental animals, are not always the best models for particular diseases. Genetic engineering of standardized mice and rats has mitigated some but not all of these concerns. The case of thalidomide, also discussed in chapter 6, demonstrated that different animals respond differently to certain drugs. In disease research, one way to mitigate differences between species is to use animal diseases that have pathologies similar to those of human diseases. For human respiratory diseases, ferrets have been recognized as a superior animal model since the 1930s. Bovine or mouse versions of pneumonia can model human pneumonia in cattle or mice, but in the end they are only models.[2] The initial choice of research animals, therefore, can make the difference between success and failure in developing vaccines and medicines. All this work builds on knowledge gained in centuries of research on human and animal bodies.

The COVID-19 pandemic is also an example of another kind of interface of human and animal science. Like many human diseases, this virus originated in animals, and infected humans can transmit it back to animals. The disease COVID-19 is caused by a virus known as SARS-CoV-2, for severe acute respiratory syndrome coronavirus-2; an outbreak of an earlier SARS coronavirus occurred in 2003/4, mainly in China. A coronavirus is a specific family of viruses that circulate mainly in animals but also occur in people. SARS-CoV-2 most likely originated in bats, but it is not clear how it made the jump from animals to humans. It is another example of a zoonosis, like anthrax and Ebola. Such recent boundary crossings as we have seen with Ebola, SARS, MERS (Middle East Respiratory Syndrome), COVID-19, and other diseases have brought new attention to the interspecies entanglements between humans and animals. Human encroachment on animal territories and climate change have brought humans into closer proximity to animals, with the transfer of diseases being one result.

According to researchers, disease research, and particularly vaccine development, is an area in which animal models remain essential. For example, a group of Dutch researchers recently stated that "the current COVID-19 crisis highlights the reality that animal research remains essential to find solutions for human and animal health" and criticized a push by Dutch and EU regulators toward less use of live animals and more use of "animal-free" alternatives.[3] Nonetheless, the search for alternatives is continuous. The "three Rs" of Russell and Burch (replacement, reduction, refinement) include replacement of animals with, for example, cell or tissue cultures; these date back to virus research in the 1930s. High-tech alternatives now also exist: toxicity testing in animals, for example, may be replaced by the use of mass spectrometry and computer-based modeling. Big data and machine learning can improve what is known as "read-across," where data on one toxic source are determined to be similar enough to data on another source to serve as a basis for assessment of its safety. The Center for Alternatives to Animal Testing at Johns Hopkins University, founded in 1981, works on these and other forms of replacement for animals. Proponents of animal-free testing methods argue that such methods (known as new approach methods, or NAMs) are ideal for the discovery of new drugs. NAMs, they argue, are faster than animal testing and may be more accurate.

Another form of replacement is of one animal for another. In the nineteenth century, as we saw in chapter 4, Marshall Hall advocated using "lower" (i.e., cold-blooded) animals such as frogs rather than mammals. In this category, the zebrafish (*Danio rerio*), a tiny freshwater tropical fish, has become a popular model organism in a large number of studies of development (particularly of the nervous system) and genetics, as well as in some forms of toxicity and disease research. The University of Oregon molecular biologist George Streisinger (1927–1984) pioneered the use of zebrafish in the 1970s, and it was the first vertebrate animal to be cloned, in 1983. As with mice and rats earlier in the twentieth century, the widespread use of zebrafish has led to a large database of genetic and genomic information, facilitating its laboratory usefulness.

The unprecedented spread of the COVID-19 pandemic has also led to an explosion of clinical trials of various potential therapies and vaccines. As we saw in chapter 3, public dread of a disease can lead to public acceptance of risk. Smallpox inoculation did create immunity against the disease, but it also carried significant risks of death and disfiguration, as well as of spreading the disease. Three hundred years later, the US National Institutes of Health issued a public-service announcement in the fall of 2020 calling for 1 million volunteers for clinical trials, framing volunteering as a civic responsibility.[4] While this was probably the biggest set of volunteers since the Polio Pioneers in 1954, modern laws and rules concerning consent made this a better-informed cohort.

As we saw in chapter 3, human and animal medicine began to diverge with the founding of veterinary schools in eighteenth-century France. Edward Jenner discovered smallpox vaccination in the 1790s owing to his intimate knowledge of the diseases of horses and cattle. Animal and human medicine occasionally merged after this time, such as in the work of Theobald Smith on cattle diseases and the transmission of bovine tuberculosis to humans. But as veterinary medicine became a distinct profession, "medicine" came to constitute human medicine, in which animals were instruments for learning about human diseases.

The boundaries between animal and human medicine have begun to break down, however, both in medicine itself and in newer research in the history of science and medicine. An interspecies history of medicine is beginning to take form that views animals as agents of their own history and not merely victims of human exploitation. In medicine, the movement known as One Health has emerged in the past two decades. One Health "is underpinned by the belief that some of the most important health threats faced today are not species specific, and consequently can only be tackled by interdisciplinary methods working across the domains of human medicine, veterinary medicine, and the life sciences."[5]

The term *One Health* first appeared in 2003, sparked by the work of the veterinarian Calvin W. Schwabe (1927–2006), whose work on public health and parasitic diseases led him to urge what he termed *One Medicine* as early as 1964. The Ebola outbreaks of the 1990s, as well as outbreaks of avian influenza (H5N1) and SARS early in the following decade, gave impetus to One Health. It has now become a global movement devoted to cross-disciplinary collaboration to ensure human, animal, and environmental health.

In closing this book with the emergence of One Health, I wish to leave readers with some optimism for the future. However, we need to acknowledge that although we are finding out more about the human and animal bodies than we have ever known before, we are not sure what to do with that knowledge, either practically or ethically. Meanwhile, emerging infectious diseases and the research required to discover new drugs and vaccines will undoubtedly use millions of animals and continue to raise questions about consent in clinical trials.

My purpose in writing this book, and particularly in undertaking a new edition, is not to resolve any of these issues or to project current moral sensibilities back on historical actors but to provide readers with the materials to understand their history and therefore to understand their present. Animal and human experimentation are unlikely to disappear in the foreseeable future, but discussion of the values that dictate policies for its conduct must come from an informed public.

Suggested Further Reading

There are few general works on animal and human experimentation. The articles in *Vivisection in Historical Perspective*, ed. Nicolaas Rupke (New York: Routledge, 1987), explore the topic from antiquity to the twentieth century, with the focus on the nineteenth century. Individual articles are noted below. Anita Guerrini, "Animal Research I: Historical Aspects," in *Bioethics*, ed. Bruce Jennings, 4th ed., vol. 1 (New York: Macmillan Reference USA, 2014), 218–23, offers a short survey. Each of the six volumes in the Bloomsbury *Cultural History of Animals*, ed. Linda Kalof and Brigitte Resl (London: Bloomsbury, 2007), includes a section on animals and science. Many antivivisectionist works contain historical material of varying value; some of these are listed below. General works in the history of biology also contain much relevant material on animal experimentation in particular. A good, if rather dry, text is T. S. Hall's *History of General Physiology*, 2 vols. (Chicago: University of Chicago Press, 1969); see also Karl Rothschuh, *History of Physiology*, trans. Günter Risse (New York: Krieger, 1973). I know of no general history of human experimentation, but see Anita Guerrini, "The Human Experimental Subject," in *A Companion to the History of Science*, ed. Bernard Lightman (New York: Wiley, 2016), 126–38, and Erika Dyck and Larry Stewart, eds., *The Uses of Humans in Experiment* (Leiden: Brill, 2016); for readers of French, Grégoire Chamayou, *Les corps vils: Expérimenter sur les êtres humains aux XVIII*[e] *et XIX*[e] *siècles* (Paris: La Découverte, 2008). Lawrence K. Altman, *Who Goes First?* (Berkeley: University of California Press, 1998), surveys self-experimentation. Reliable guides to current science may be found in the news section of the journals *Science* and *Nature.*

1. Bodies of Evidence

Richard Sorabji's *Animal Minds and Human Morals* (Ithaca, NY: Cornell University Press, 1993) explores ancient ideas about animals. John Passmore's "The Treatment of Animals," *Journal of the History of Ideas* 36 (1975): 195–218, is a useful short survey of ancient ideas. The Alexandrians receive a thorough treatment in Heinrich von Staden, *Herophilus* (Cambridge: Cambridge University

Press, 1989). Essential to understanding Aristotle is G. E. R. Lloyd, *Aristotle* (Cambridge: Cambridge University Press, 1968); for Galen, see Susan P. Mattern, *The Prince of Medicine: Galen in the Roman Empire* (New York: Oxford University Press, 2013). Nancy Siraisi's *Medieval and Renaissance Medicine* (Chicago: University of Chicago Press, 1990) is an excellent survey of ideas and practices. Relevant to chapters 1–3 is A. H. Maehle and U. Tröhler, "Animal Experimentation from Antiquity to the End of the Eighteenth Century," in *Vivisection in Historical Perspective*, ed. Nicolaas Rupke (New York: Routledge, 1987).

2. Animals, Machines, and Morals

There is a vast historical literature on Harvey. Gweneth Whitteridge's *William Harvey and the Circulation of the Blood* (London: Macdonald, 1971) offers a thorough account of Harvey's discovery, emphasizing his experiments; see also Andrew Wear's introduction to William Harvey, *The Circulation of the Blood, and Other Writings*, trans. K. Franklin (London: Dent, Everyman's Library, 1990). Jole Shackelford's *William Harvey and the Mechanics of the Heart* (New York: Oxford University Press, 2003) is an excellent short survey. The Royal College of Physicians of London produced a video on Harvey and circulation in 1970 that is available on YouTube. On Cartesianism and animals, the major source remains Leonora Cohen Rosenfield, *From Beast-Machine to Man-Machine* (1940; reprint, New York: Octagon, 1968). On mechanical physiology, see R. G. Frank Jr., *Harvey and the Oxford Physiologists* (Berkeley: University of California Press, 1980); T. M. Brown, *The Mechanical Philosophy and the "Animal Oeconomy"* (New York: Arno, 1981); Anita Guerrini, "The Ethics of Animal Experimentation in Seventeenth-Century England," *Journal of the History of Ideas* 50 (1989): 391–407; Domenico Bertoloni Meli, *Mechanism, Experiment, Disease* (Baltimore: Johns Hopkins University Press, 2011); Domenico Bertoloni Meli and Anita Guerrini, eds., "Experimenting with Animals in the Early Modern Era," special issue, of *Journal of the History of Biology* 46, no. 2 (May 2013); and Bertoloni Meli, ed., *Marcello Malpighi, Physician and Anatomist* (Florence: L. Olschki, 1997).

3. Disrupting God's Plan

Still a major contribution to understanding the inoculation controversy is Genevieve Miller, *The Introduction of Smallpox Inoculation in England and France* (Philadelphia: University of Pennsylvania Press, 1957). A broader history of smallpox is Donald R. Hopkins, *Princes and Peasants: Smallpox in History* (Chicago: University of Chicago Press, 1983).

On Jenner, see Derrick Baxby, *Jenner's Smallpox Vaccine: The Riddle of Vaccinia Virus and Its Origin* (London: Heinemann Educational Books, 1981). On the modern eradication campaign, the main source is Frank Fenner, Donald A. Henderson, Isao Arita, Zdenek Jezek, and Ivan Danilovich Ladnyi, *Smallpox and Its Eradication* (Geneva: World Health Organization, 1988); a more accessible account is Bob H. Reinhardt, *The End of a Global Pox* (Chapel Hill: University of North Carolina Press, 2015). Isobel Grundy's *Lady Mary Wortley Montagu* (Oxford: Clarendon, 1999) contains much detail on Montagu's role in inoculation. The role of statistics is well explained in Andrea Rusnock, "The Weight of Evidence and the Burden of Authority: Case Histories, Medical Statistics and Smallpox Inoculation," in *Medicine in the Enlightenment*, ed. Roy Porter (Amsterdam: Rodopi, 1995). On Hales, see D. G. C. Allan and R. E. Schofield, *Stephen Hales: Scientist and Philanthropist* (London: Scolar Press, 1980).

4. Cruelty and Kindness

The history of experimental physiology in the nineteenth century has a vast bibliography. John Lesch's *Science and Medicine in France: The Emergence of Experimental Physiology, 1790–1855* (Cambridge, MA: Harvard University Press, 1984) discusses the period before Bernard; F. L. Holmes's *Claude Bernard and Animal Chemistry* (Cambridge, MA: Harvard University Press, 1974) closely studies the earlier years of Bernard's career. Also important are Joseph Schiller's *Physiology and Classification* (Paris: Maloine, 1980) and his "Claude Bernard and Vivisection," *Journal of the History of Medicine and Allied Sciences* 22 (1967): 246–60. The standard biography of Bernard remains J. M. D. Olmsted and E. H. Olmsted, *Claude Bernard and the Experimental Method in Medicine* (New York: H. Schuman, 1952). On Marshall Hall, see Diana Manuel's *Marshall Hall (1790–1857): Science and Medicine in Early Victorian Society* (Amsterdam: Rodopi, 1996); and Manuel, "Marshall Hall (1790–1857): Vivisection and the Development of Experimental Physiology," in *Vivisection in Historical Perspective*, ed. Nicolaas Rupke (New York: Routledge, 1987), 78–104. The Anatomy Act of 1832 is the subject of Ruth Richardson's brilliant *Death, Dissection, and the Destitute* (London: Routledge, 1988). On Beaumont, see Ronald Numbers, "William Beaumont and the Ethics of Human Experimentation," *Journal of the History of Biology* 12 (1979): 113–35; and Alexa Green, "Working Ethics: William Beaumont, Alexis St. Martin, and Medical Research in Antebellum America," *Bulletin of the History of Medicine* 84 (2010), 193–216; on the use of slaves, see Todd L. Savitt, "The Use of Blacks for Medical Experimentation and Demonstration in the Old South," *Journal of Southern History* 48 (1982):

331–48. Martin Pernick treats the introduction of anesthesia in *A Calculus of Suffering* (New York: Columbia University Press, 1985); see also Stephanie Snow, *Operations without Pain: The Practice and Science of Anaesthesia in Victorian Britain* (Basingstoke, Hants.: Palgrave Macmillan 2006). Harriet Ritvo's *The Animal Estate: The English and Other Creatures in the Victorian Age* (Cambridge, MA: Harvard University Press, 1987) looks at the relationships between animals and the Victorians; for the French, see Kathleen Kete, *The Beast in the Boudoir: Petkeeping in Nineteenth-Century Paris* (Berkeley: University of California Press, 1994), which is broader than its title indicates. The most important source for the nineteenth-century antivivisection movement remains Richard French, *Antivivisection and Medical Science in Victorian Society* (Princeton, NJ: Princeton University Press, 1975); see also James Turner, *Reckoning with the Beast: Animals, Pain, and Humanity in the Victorian Mind* (Baltimore: Johns Hopkins University Press, 1982), and Lloyd Stevenson, "Science Down the Drain," *Bulletin of the History of Medicine* 29 (1955): 1–26. Coral Lansbury, *The Old Brown Dog* (Madison: University of Wisconsin Press, 1986), and Lori Williamson, *Power and Protest: Frances Power Cobbe and Victorian Society* (London: Rivers Oram, 1998), discuss the role of women. Susan Lederer's "Controversy in America, 1880–1914" and Mary Ann Elston's "Women and Antivivisection in Victorian England" in the Rupke volume are also relevant. For the antivivisectionist point of view, see John Vyvyan's *In Pity and in Anger* (London: Michael Joseph, 1969) and *The Dark Face of Science* (London: Michael Joseph, 1971); and Richard D. Ryder, *Victims of Science: The Use of Animals in Research* (London: Davis-Poynter, 1975).

5. The Microbe Hunters

C. E. A. Winslow's *The Conquest of Epidemic Disease* (Princeton, NJ: Princeton University Press, 1943) remains a classic account of the bacteriological revolution, especially useful for its long historical perspective. Paul de Kruif's *Microbe Hunters* (New York: Harcourt, Brace, 1926) conveys the excitement of research by one who participated in it; Sinclair Lewis's *Arrowsmith* (1925, many editions) is a fictional account of many of the same events. Both served as effective propaganda for science in the 1920s. Gerald Geison's *The Private Science of Louis Pasteur* (Princeton, NJ: Princeton University Press, 1995), based on a close reading of Pasteur's laboratory notebooks, reveals a less heroic but more human Pasteur; heroic stature is restored in Patrice Debré, *Louis Pasteur*, trans. E. Forster (Baltimore: Johns Hopkins University Press, 1998). For Koch, see Thomas Brock, *Robert Koch: A Life in Medicine and Bacteriology* (Madison, WI: Science Tech, 1988). Elie Metchnikoff, ed., *The Founders of Modern Medicine:*

Pasteur, Koch, Lister, trans. D. Berger (New York: Walden, 1939), reprints original papers. Harry F. Dowling's *Fighting Infection: Conquests of the Twentieth Century* (Cambridge, MA: Harvard University Press, 1977) is a reliable guide to the development of antibiotics; see also John Parascandola, ed., *The History of Antibiotics: A Symposium* (Madison, WI: American Institute for the History of Pharmacy, 1980), and Harry M. Marks, *The Progress of Experiment: Science and Therapeutic Reform in the United States, 1900–1990* (New York: Cambridge University Press, 1997). Susan E. Lederer, "Political Animals: The Shaping of Biomedical Research Literature in Twentieth-Century America," *Isis* 83 (1992): 61–79, discusses scientists' response to the perceived threat from antivivisectionists. Somewhat outside the scope of this book but offering a valuable perspective on the transfer of Western science to colonized nations in the nineteenth century is Pratik Chakrabarti, "Beasts of Burden: Animals and Laboratory Research in Colonial India," *History of Science* 48 (2010): 125–51.

6. Polio and Primates

A prizewinning account of primate research, Deborah Blum's *The Monkey Wars* (New York: Oxford University Press, 1994), offers a balanced view. A feminist-postmodern critique of primate research is offered in Donna Haraway, *Primate Visions* (New York: Routledge, 1990). The early years of polio research are detailed in Naomi Rogers, *Dirt and Disease: Polio before FDR* (New Brunswick, NJ: Rutgers University Press, 1992). An authoritative look at twentieth-century polio is David Oshinsky's *Polio: An American Story* (New York: Oxford University Press, 2005); see also the physician Paul Offit's *The Cutter Incident: How America's First Polio Vaccine Led to the Growing Vaccine Crisis* (New Haven, CT: Yale University Press, 2005). On the 1954 field trials, see Allan M. Brandt, "Polio, Politics, Publicity, and Duplicity: Ethical Aspects in the Development of the Salk Vaccine," *International Journal of Health Services* 8 (1978): 257–70; and Marcia Meldrum, "'A Calculated Risk': The Salk Polio Vaccine Field Trials of 1954," *British Medical Journal* 317 (1998): 1233–36. Harry Harlow's famous presidential address to the American Psychological Association was published as "The Nature of Love," *American Psychologist* 13 (1958): 673–85. For an analysis of Harlow by one of his former students, see John P. Gluck, "Harry Harlow and Animal Research: Reflection on the Ethical Paradox," *Ethics and Behavior* 7, no. 2 (1997): 149–61; and Gluck, *Voracious Science and Vulnerable Animals: A Primate Scientist's Ethical Journey* (Chicago: University of Chicago Press, 2016). Deborah Blum examines Harlow in *Love at Goon Park: Harry Harlow and the Science of Affection* (New York: Berkley, 2002). Arnold Arluke summarized his work on animal workers in "Trapped in a Guilt Cage," *New*

Scientist 134 (4 April 1992): 33–35. Peter Singer's *Animal Liberation*, 2d ed. (New York: New York Review of Books, 1990), and Tom Regan's *The Case for Animal Rights* (Berkeley: University of California Press, 1983) remain the touchstones of the movement for animal protection. The Great Ape Project is described in Paola Cavalieri and Peter Singer, eds., *The Great Ape Project* (London: Fourth Estate, 1993). Lori Gruen documents the last one thousand chimpanzees in American research facilities in https://last1000chimps.com/. On Pavlov, see Daniel Todes, *Ivan Pavlov: A Russian Life in Science* (New York: Oxford University Press, 2014).

7. From Nuremberg to CRISPR

On Nazi medical ideology, see Robert N. Proctor, *Racial Hygiene* (Cambridge, MA: Harvard University Press, 1988); on some of the ethical implications, see Arthur L. Caplan, ed., *When Medicine Went Mad: Bioethics and the Holocaust* (Totowa, NJ: Humana, 1992), and George Annas and Michael Grodin, eds., *The Nazi Doctors and the Nuremberg Code* (New York: Oxford University Press, 1992). Sheldon Harris detailed Japanese wartime experiments in *Factories of Death: Japanese Biological Warfare, 1932–45, and the American Cover-Up* (New York: Routledge, 1994). Andrew C. Ivy's justification for research on prisoners is included in "The History and Ethics of the Use of Human Subjects in Medical Experiments," *Science* 108 (1948): 1–5. One of Ivy's paradigm cases for experimentation on prisoners is examined in Nathaniel Comfort, "The Prisoner as Model Organism: Malaria Research at Stateville Penitentiary," *Studies in History and Philosophy of Biological and Biomedical Sciences* 40 (2009): 190–203. Allen M. Hornblum, *Acres of Skin: Human Experiments at Holmesburg Prison* (New York: Routledge, 1998), details decades of experiments on prisoners at a Philadelphia penitentiary. On the development of protections for human subjects, see Jay Katz, Alexander Capron, and Eleanor Swift Glass, *Experimentation with Human Beings* (New York: Russell Sage Foundation, 1972); David Rothman, *Strangers at the Bedside: A History of How Law and Bioethics Transformed Medical Decision Making* (New York: Basic Books, 1991); and Ruth R. Faden and Tom L. Beauchamp, *A History and Theory of Informed Consent* (New York: Oxford University Press, 1986). The text of the Helsinki Declaration and subsequent amendments can be found at https://www.wma.net/what-we-do/medical-ethics/declaration-of-helsinki/. James H. Jones analyzed the Tuskegee syphilis study in *Bad Blood*, 2d ed. (New York: Free Press, 1991); more recently, Susan Reverby's *Examining Tuskegee: The Infamous Syphilis Study and Its Legacy* (Chapel Hill: University of North Carolina Press, 2009) reexamines the study and its legacy of myth and misinformation. On the Guatemala case, see Susan

Reverby, "Ethical Failures and History Lessons: The U.S. Public Health Service Research Studies in Tuskegee and Guatemala," *Public Health Reviews* 34 (2012): 1–18. Abuses in the 1960s were revealed by Henry Beecher in "Ethics and Clinical Research," *New England Journal of Medicine* 274 (1966): 1354–60, and Maurice H. Pappworth, *Human Guinea Pigs: Experimentation on Man* (London: Routledge & Kegan Paul, 1967); Beecher's article is reprinted with a historical introduction in Jon Harkness, Susan E. Lederer, and Daniel Wikler, "Laying Ethical Foundations for Clinical Research," *Bulletin of the World Health Organization* 79 (2001): 365–72. Susan E. Lederer, *Subjected to Science* (Baltimore: Johns Hopkins University Press, 1995), discusses human experimentation in the United States before World War II. Radiation experiments are documented in United States, Advisory Committee on Human Radiation Experiments, *Final Report* (Washington, DC: GPO, 1995).

The use of children in medical experimentation after World War II is documented in Allen M. Hornblum, Judith L. Newman, and Gregory J. Dober, *Against Their Will: The Secret History of Medical Experimentation on Children in Cold War America* (New York: St. Martin's Press, 2013). Continued abuses of human subjects are explored in Sonia Shah, *The Body Hunters: Testing New Drugs on the World's Poorest Patients* (New York: New Press, 2006); and Roberto Abadie, *The Professional Guinea Pig: Big Pharma and the Risky World of Human Subjects* (Durham, NC: Duke University Press, 2010).

"Concentration Camps for Dogs," in *Life*, 4 February 1966, 22–29, ignited the campaign for the 1966 US Animal Welfare Act. See also W. L. M. Russell and R. L. Burch, *The Principles of Humane Experimental Technique* (London: Methuen, 1959). Annual reports of animal usage under the AWA are found at www.aphis.usda.gov. On the experimental mouse, see Karen Rader, *Making Mice* (Princeton, NJ: Princeton University Press, 2004); on rats, Anita Guerrini, "A Tale of Two Rats," in *Nature Remade: Engineering Life, Envisioning Worlds*, ed. Luis Campos, Michael R. Dietrich, Tiago Saraiva, and Christian Young (Chicago: University of Chicago Press, 2021), 31–43. Manuel Berdoy's "The Laboratory Rat: A Natural History" can be viewed on YouTube. An excellent account of contemporary lab animal welfare is Larry Carbone, *What Animals Want: Expertise and Advocacy in Laboratory Animal Welfare Policy* (New York: Oxford University Press, 2004).

On genetics, see Nathaniel Comfort, *The Science of Human Perfection* (Baltimore: Johns Hopkins University Press, 2012); and Rebecca Skloot, *The Immortal Life of Henrietta Lacks* (New York: Broadway Books, 2011). A helpful guide to modern genetics for nonscientists can be found at https://medlineplus.gov/genetics/.

The Oxford Handbook of Animal Ethics, ed. Tom L. Beauchamp and R. L. Frey (New York: Oxford University Press, 2014), offers a comprehensive survey by leading philosophers. The Animal Welfare Information Center of the US Department of Agriculture (https://www.nal.usda.gov/awic) includes texts of the relevant laws. A comprehensive overview of animal research laws and standards across the globe is in a thematic issue of *ILAR Journal* 57, no. 3 (2016), "International Laws, Regulations, and Guidelines for Animal Research," introduced by Mary Ann Vasbinder and Paul Locke. On food animals and veterinary science, see Susanne Freidberg, *Fresh: A Perishable History* (Cambridge, MA: Harvard University Press, 2009), and Susan D. Jones, *Valuing Animals* (Baltimore: Johns Hopkins University Press, 2003); the 2015 exposé of USMARC is Michael Moss, "U.S. Research Lab Lets Livestock Suffer in Quest for Profit," *New York Times*, 19 January 2015, https://nyti.ms/1AEPr4J.

Robert Kohler explores the tension between field and lab in *Landscapes and Labscapes: Exploring the Lab-Field Border in Biology* (Chicago: University of Chicago Press, 2002). On rewilding, see C. Josh Donlan et al., "Re-wilding North America," *Nature* 436 (2005): 913–14; on Oostvaardersplassen, see Bert Theunissen, "The Oostvaardersplassen Fiasco," *Isis* 110 (2019): 341–45. Issues surrounding de-extinction are explored in Dolly Jørgensen, "Reintroduction and De-extinction," *Bioscience* 63 (2013): 719–20; and Lesley Evans Ogden, "Extinction Is Forever . . . or Is It?," *Bioscience* 64 (2014): 469–75.

On the Jesse Gelsinger case, see Meir Rinde, "The Death of Jesse Gelsinger, 20 Years Later," *Distillations*, 4 June 2019, https://www.sciencehistory.org/distillations/the-death-of-jesse-gelsinger-20-years-later. On disability, see David Wasserman, Adrienne Asch, Jeffrey Blustein, and Daniel Putnam, "Disability: Definitions, Models, Experience," in *Stanford Encyclopedia of Philosophy*, https://plato.stanford.edu/archives/sum2016/entries/disability/. George Estreich, *Fables and Futures: Biotechnology, Disability, and the Stories We Tell Ourselves* (Cambridge, MA: MIT Press, 2019), offers a thoughtful analysis of the issues surrounding genetic engineering.

Conclusion

On One Health and its history, see Abigail Woods, Michael Bresalier, Angela Cassidy, and Rachel Mason Dentinger, *Animals and the Shaping of Modern Medicine: One Health and Its Histories* (London: Palgrave Macmillan, 2018).

Notes

Introduction

1. Jon Cohen, "From Mice to Monkeys, Animals Studied for Coronavirus Answers," *Science* 368, no. 6488 (2020): 221–22.

2. Evan Callaway, "Hundreds of People Volunteer to Be Infected with Coronavirus," *Nature*, 22 April 2020, https://www.nature.com/articles/d41586=020–01179-x.

3. Susan M. Reverby, "Ethical Failures and History Lessons: The U.S. Public Health Service Research Studies in Tuskegee and Guatemala," *Public Health Reviews* 34 (2012): 1–18, at 3.

1. Bodies of Evidence

1. Heinrich von Staden, *Herophilus* (Cambridge: Cambridge University Press, 1989), 141.

2. Aristotle, *De Partibus Animalium (On the Parts of Animals)*, trans. J. Balme (Oxford: Clarendon, 1972), 645a.15–20, 20–25.

3. Galen, *On Anatomical Procedures*, trans. Charles Singer (London: Oxford University Press, 1956), 33–34.

4. Galen, *On Anatomical Procedures, the Later Books*, trans. W. L. H. Duckworth (Cambridge: Cambridge University Press, 1962), 15.

5. Galen, *On Anatomical Procedures* (1956), 192.

6. Gen. 1:28. All references are to the New Revised Standard Version.

7. Eccles. 3:19.

8. Matt. 23:37; Luke 15:3–7.

9. Augustine, *The City of God*, trans. Henry Bettenson (London: Penguin, 1984), 19.15.

10. Augustine, *The Catholic and Manichaean Ways of Life*, trans. D. A. Gallagher and L. J. Gallagher (Washington, DC: Catholic University of America Press, 1966), 102, 105.

2. Animals, Machines, and Morals

1. Caspar Hofmann to William Harvey, May 1636, trans. Gweneth Whitteridge, in Whitteridge, *William Harvey and the Circulation of the Blood* (London: Macdonald, 1971), 241.

2. William Harvey, *Anatomical Exercises on the Generation of Animals* (1651), in *The Works of William Harvey, M.D.*, trans. Robert Willis (London: Sydenham Society, 1847), 151. Willis adds to the title "Anatomical," which is not in the Latin.

3. Peter Singer, *Animal Liberation*, 2d ed. (1990; reprint, New York: Avon, 1991), 200; Tom Regan, *The Case for Animal Rights* (Berkeley: University of California Press, 1983), 5.

4. Institute of Medical Ethics, *Lives in the Balance*, ed. J. A. Smith and K. M. Boyd (Oxford: Oxford University Press, 1991), 300.

5. Johann Conrad Brunner, *Experimenta nova circa pancreas* (Amsterdam: Henricus Wetstein, 1683), 57.

6. *The Diary of John Evelyn*, ed. E. S. de Beer, 6 vols. (Oxford: Clarendon, 1955), 3:497–98.

7. Richard Lower, "The Method Observed in Transfusing the Blood out of One Animal into Another," *Philosophical Transactions* 2 (1666): 353–58.

8. Edmund King to Robert Boyle, 25 November 1667, in *The Correspondence of Robert Boyle*, ed. Michael Hunter, Antonio Clericuzio, and Lawrence M. Principe, 7 vols. (London: Pickering & Chatto, 2001), 3: 366–67.

9. *The Correspondence of Marcello Malpighi*, ed. H. B. Adelmann, 5 vols. (Ithaca: Cornell University Press, 1975), vol. 1.

10. Robert Boyle, "New Pneumatical Experiments about Respiration," *Philosophical Transactions* 5 (1670): 2044.

11. Carlo Fracassati, "An Account of Some Experiments of Injecting Liquors into the Veins of Animals," *Philosophical Transactions* 2 (1667): 490.

12. Niels Stensen to Thomas Bartholin, 1661, quoted in Frederik Ruysch, *Dilucidatio valvularum in vasis lymphaticis et lacteis* (1665), ed. A. M. Luyendijk-Elshout (Nieuwkoop, Netherlands: B. de Graaf, 1964), 36.

13. John Ray, "De animalibus in genere," in *Synopsis methodica animalium quadrupedum et serpentini generis* (London, 1693), 12, translated in Charles Raven, *John Ray, Naturalist* (Cambridge: Cambridge University Press, 1942), 375.

3. Disrupting God's Plan

1. Quoted in Genevieve Miller, *The Introduction of Smallpox Inoculation in England and France* (Philadelphia: University of Pennsylvania Press, 1957), 56.

2. Charles Maitland to Sir Hans Sloane, n.d., ca. 1721, British Library, Sloane MSS 4034, fol. 18.

3. William Douglass, *Abuses and Scandals* (London, 1722), introduction.

4. William Wagstaffe, *A Letter to Dr Freind; Shewing the Danger and Uncertainty of Inoculating the Small Pox* (London, 1722), 5–6.

5. Grégoire Chamayou, *Les corps vils* (Paris: La Découverte, 2008), 7.

6. Stephen Hales, *Vegetable Staticks* (1727), ed. M. A. Hoskins (London: Macdonald, 1969), xxxi.

7. Joseph Spence, *Observations, Anecdotes, and Characters of Books and Men: Collected from the Conversation of Mr. Pope* (London: W. B. Carpenter; Edinburgh: Archibald Constable, 1820), 293–94; Thomas Twining, "The Boat" (ca. 1740), quoted in Marjorie Hope Nicolson and G. S. Rousseau, *This Long Disease, My Life: Alexander Pope and the Sciences* (Princeton, NJ: Princeton University Press, 1968), 104–5.

8. Samuel Johnson, *The Idler* no. 17 (5 August 1758).

9. Albrecht von Haller, *A Treatise on the Sensible and Irritable Parts of Animals* (London, 1755), 1.

10. Junius, "The Air-Pump," *Gentleman's Magazine* 10 (1740): 194; Christlob Mylius, "Untersuchung ob man die Thiere, um physiologischer Versuche willen, lebendig eröffnen dürfe" [Investigating whether animals should be opened up alive for the sake of physiological experiments], *Belustigungen des Verstandes und des Witzes*, April 1745, 325–40.

11. Jeremy Bentham, *An Introduction to the Principles of Morals and Legislation* (London: T. Payne, 1789), ccix.

12. Edward Jenner, *An Inquiry into the Causes and Effects of the Variolae Vaccinae, a Disease Discovered in Some of the Western Counties of England, particularly Gloucestershire, and known by the name of the Cow Pox* (London: printed for the author, 1798), 2–3, 6.

13. Derrick Baxby, "The Origins of Vaccinia Virus," *Journal of Infectious Diseases* 136, no. 3 (1977): 453–55, at 454; Kai Kupferschmidt, "How Canadian Researchers Reconstituted an Extinct Poxvirus for $100,000 Using Mail-Order DNA," *Science*, 7 July 2017, doi:10.1126/science.aan7069.

4. Cruelty and Kindness

1. "Bear-Baiting Prevention Bill," 24 February 1825, *Hansard Parliamentary Debates* (Commons), n.s. 12, cols.657–59.

2. J. J. C. Legallois, *Expériences sur la principe de la vie* (Paris, 1812), xxiii, quoted in J. E. Lesch, *Science and Medicine in France: The Emergence of Experimental Physiology, 1790–1855* (Cambridge, MA: Harvard University Press, 1984), 89.

3. "Cruelty to Animals Bill," 11 March 1825, *Hansard Parliamentary Debates* (Commons), n.s. 12, cols. 1010–12.

4. Marshall Hall, *A Critical and Experimental Essay on the Circulation* (London: Seeley & Burnside, 1831), 1–11 (introduction).

5. Fanny Burney d'Arblay to Esther Burney, 22 March 1812, in *The Journals and Letters of Fanny Burney*, ed. Joyce Hemlow, vol. 6 (London: Oxford University Press, 1975), 596–615.

6. François Magendie, *Elementary Compendium of Physiology: For the Use of Students*, trans. E. Milligan (Edinburgh: John Carfrae & Son, 1831), 98.

7. George Hoggan, *Anaesthetics and Lower Animals* (London, 1875), quoted in Richard French, *Antivivisection and Medical Science in Victorian Society* (Princeton, NJ: Princeton University Press, 1975), 68.

8. Claude Bernard, *Principes de médecine expérimentale* (Paris, 1947), 285, trans. and quoted in William Coleman, "The Cognitive Basis of the Discipline: Claude Bernard on Physiology," *Isis* 76 (1985): 56.

9. Claude Bernard, *An Introduction to the Study of Experimental Medicine* (1865), trans. Henry Copley Greene (1927; reprint, New York: Dover, 1957), 101–2.

10. Bernard, *Introduction*, 99.

11. Bernard, *Introduction*, 102.

12. Morris Fishbein, *The Medical Follies* (New York: Boni & Liveright, 1925), 154.

5. The Microbe Hunters

1. Louis Pasteur, "Prevention of Rabies" (1885), in *The Founders of Modern Medicine: Pasteur, Koch, Lister*, ed. Elie Metchnikoff, trans. D. Berger (New York: Walden, 1939), 383.

2. George Bernard Shaw, *The Doctor's Dilemma* (1913; reprint, Harmondsworth: Penguin, 1954), 29.

3. Charles Dickens, *American Notes for General Circulation* (New York: Harper & Brothers, 1842), 39.

4. *New York Times*, 20 November 1938, 5.

5. William Henry Welch, dean of the School of Public Health at the Johns Hopkins University, in 1926, quoted in Susan E. Lederer, "Political Animals: The Shaping of Biomedical Research Literature in Twentieth-Century America," *Isis* 83 (1992): 62.

6. Lederer, "Political Animals," 69 (quoting Walter Cannon), 61.

6. Polio and Primates

1. Naomi Rogers, *Dirt and Disease: Polio before FDR* (New Brunswick, NJ: Rutgers University Press, 1992), 77.

2. Isabel M. Morgan, "Immunization of Monkeys with Formalin-Inactivated Poliomyelitis Viruses," *American Journal of Hygiene* 48 (1948): 394–406.

3. Jonas Salk, "Principles of Immunization as Applied to Poliomyelitis and Influenza," *American Journal of Public Health* 43 (November 1953): 1384–98.

4. "The End of Polio Is in Sight at Last," *Life*, 27 October 1952, 115–16.

5. "Closing In on Polio," *Time*, 29 March 1954, 56–66.

6. Marcia Meldrum, "'A Calculated Risk': The Salk Polio Vaccine Field Trials of 1954," *British Medical Journal* 317 (1998): 1233–36, at 1234.

7. K. Stratton, D. A. Almario, and M. C. McCormick, eds., *Immunization Safety Review: SV40 Contamination of Polio Vaccine and Cancer* (Washington, DC: National Academies Press, 2002), 59.

8. Richard Burian, "How the Choice of Experimental Organism Matters: Epistemological Reflections on an Aspect of Biological Practice," *Journal of the History of Biology* 26 (1993): 351–67, at 352.

9. Edward Tyson, *Ourang-outang, sive Homo Sylvestris: or, the Anatomy of a Pygmie Compared with that of a Monkey, and Ape, and a Man* (London: Thomas Bennet & Daniel Brown, 1699), 2.

10. John B. Watson, *Psychological Care of Infant and Child* (New York: Norton, 1928), 87.

11. John Gluck, *Voracious Science and Vulnerable Animals: A Primate Scientist's Ethical Journey* (Chicago: University of Chicago Press, 2016); Arnold Arluke, "Trapped in a Guilt Cage," *New Scientist* 134 (4 April 1992): 33–35.

12. Craig B. Stanford, *The Hunting Apes: Meat Eating and the Origins of Human Behavior* (Princeton, NJ: Princeton University Press, 1999), 5.

13. See www.greatapeproject.uk, accessed 23 August 2020.

14. New Zealand Legislation, *Animal Welfare Act 1999*, http://www.legislation.govt.nz/act/public/1999/0142/latest/whole.html#DLM49664; EUR-Lex, Access to European Union law, *Directive 2010/63/EU of the European Parliament and of the Council of 22 September 2010 on the protection of animals used for scientific purposes*, sec. 18, https://eur-lex.europa.eu/legal-content/EN/TXT/?uri=uriserv:OJ.L_.2010.276.01.0033.01.ENG&toc=OJ:L:2010:276:TOC; *Bundesrecht konsolidiert: Gesamte Rechtsvorschrift für Tierversuchsgesetz 2012*, version of 24 August 2020, https://www.ris.bka.gv.at/GeltendeFassung.wxe?Abfrage=Bundesnormen&Gesetzesnummer=20008142.

15. Jocelyn Kaiser, "An End to US Chimp Research," *Science* 350, no. 6284 (27 November 2015): 1013.

7. From Nuremberg to CRISPR

1. Jay Katz, "The Regulation of Human Experimentation in the United States: A Personal Odyssey," *IRB: Ethics & Human Research* 9, no. 1 (1987): 1–6, at 2.

2. Robert Proctor, "The Destruction of 'Lives Not Worth Living,'" in *Deviant Bodies*, ed. Jennifer Terry and Jacqueline Urla (Bloomington: Indiana University Press, 1995), 170–96, at 172.

3. Arthur L. Caplan, "How Did Medicine Go So Wrong?," in *When Medicine Went Mad: Bioethics and the Holocaust*, ed. Caplan (Totowa, NJ: Humana, 1992), 57.

4. Katz, "Regulation of Human Experimentation," 2.

5. Jon Harkness, "Nuremberg and the Issue of Wartime Experiments on US Prisoners: The Green Committee," *JAMA* 276 (1996): 1672–75, at 1673.

6. Susan M. Reverby, *Examining Tuskegee: The Infamous Syphilis Study and Its Legacy* (Chapel Hill: University of North Carolina Press, 2009), 51.

7. World Medical Association, "Declaration of Helsinki: Medical Research Involving Human Subjects, June 1964," https://www.wma.net/what-we-do/medical-ethics/declaration-of-helsinki/.

8. Maurice H. Pappworth, *Human Guinea Pigs: Experimentation on Man* (London: Routledge & Kegan Paul, 1967), 185–87.

9. *The Animal Welfare Act—Public Law 89–544 Act of August 24, 1966*, https://www.nal.usda.gov/awic/animal-welfare-act-public-law-89–544-act-august-24–1966.

10. Judith Hampson, "Legislation and the Changing Consensus," in *Animal Experimentation: The Consensus Changes*, ed. Gill Langley (New York: Chapman & Hall, 1989), 220.

11. Andrea L. Beach and David E. Wright, "A Short History of Regulations Protecting Human Subjects Research," *Research Integrity* 4, no. 1a (Spring 2000): 5.

12. Theobald Smith and F. L. Kilborne, *Investigations into the Nature, Causation, and Prevention of Texas or Southern Cattle Fever*, US Department of Agriculture, Bureau of Animal Industry, Bulletin no. 1 (Washington, DC: GPO, 1893).

13. See https://www.ars.usda.gov/plains-area/clay-center-ne/marc/, accessed 5 September 2020.

14. "Policies. Regulation of Agricultural Animals, Policy #17, USDA,"

https://www.nal.usda.gov/sites/default/files/Policy17_0.pdf accessed 5 September 2020.

15. American Society of Animal Science, "USDA releases final report of USMARC audit," 22 December 2016, https://www.asas.org/taking-stock/blog-post/taking-stock/2016/12/22/usda-releases-final-report-of-usmarc-audit. The final report appears to be unavailable on the USDA website.

16. Keith R. Benson, "The Emergence of Ecology from Natural History," *Endeavour* 24, no. 2 (2000): 59–62.

17. C. Josh Donlan, "De-extinction in a Crisis Discipline," *Frontiers of Biogeography* 6 (2014): 25–28.

18. Nathaniel Comfort, *The Science of Human Perfection* (Baltimore: Johns Hopkins University Press, 2012), 200–201.

19. "United Nations Declaration on Human Cloning," United Nations General Assembly, Resolution 59/80, adopted 8 March 2005.

20. Stan J. J. Brouns, "A Swiss Army Knife of Immunity," *Science* 337 (17 August 2012): 808–9.

21. Jennifer Doudna and Samuel H. Sternberg, *A Crack in Creation: Gene Editing and the Unthinkable Power to Control Evolution* (New York: Houghton Mifflin Harcourt, 2017), 181.

Conclusion

1. See Anita Guerrini, "Animals and the COVID Vaccine," *Endeavour* 45, no. 3 (2021), https://doi.org/10.1016/j.endeavour.2021.100779.

2. Volker Gerdts, Sylvia van Drunen, Littel-van den Hurk, Philip J. Griebel, and Lorne A. Babiuk, "Use of Animal Models in the Development of Human Vaccines," *Future Microbiology* 2, no. 6 (2007): 667–75, at 669.

3. Lisa Genzel et al., "How the COVID-19 Pandemic Highlights the Necessity of Animal Research," *Current Biology* 30 (2020): R1014–R1018, at R1017.

4. Clifton Leaf, "We Need a Million More Volunteers for COVID Vaccine and Drug Trials: Can Harrison Ford Convince You to Sign Up?," *Fortune*, 27 September 2020, https://fortune.com/2020/09/27/million-volunteers-covid-vaccine-trials-harrison-ford-convince-sign-up/.

5. Abigail Woods, Michael Bresalier, Angela Cassidy, and Rachel Mason Dentinger, "Introduction: Centring Animals within Medical History," in Woods, Bresalier, Cassidy, and Mason Dentinger, *Animals and the Shaping of Modern Medicine: One Health and its Histories* (London: Palgrave Macmillan, 2018), 1–26, at 14.

Index